Introduction to
topological dynamics

W0037636

Introduction to topological dynamics

K. S. SIBIRSKY

Academy of Sciences of the
Moldavian Soviet Socialist Republic
Kishinyov, USSR

Translated by

Leo F. Boron

University of Idaho, Moscow, USA

NOORDHOFF INTERNATIONAL PUBLISHING
LEYDEN

ISBN-13:978-94-010-2310-8   e-ISBN-13:978-94-010-2308-5
DOI:10.1007/978-94-010-2308-5

Original edition published in 1970 under the title
'*Vvedenie v topologicheskuyu dinamiku*'
in Kishinyov.

CONTENTS

Contents

Chapter IV. Minimal sets and recurrent motions

Chapter V. Almost periodic motions. Lyapunov stability

Chapter VI. Generalized theory of dynamical systems

FOREWORD TO THE AMERICAN EDITION

The reader should note that the important items throughout each section of this book are numbered consecutively, *e.g.*, the first definition of Section 16 is labeled Definition 1.16, the second theorem of Section 16 is labeled Theorem 2.16, the sixth corollary of Section 25 is labeled Corollary 6.25, and so forth. This numbering system does not make use of the chapter numbers.

This American edition differs from the Russian original only in that typographical and other similar small errors have been corrected and the literature list extended to include important literature that appeared after the Russian edition came off the press in 1970.

The volume and page and/or review number of Mathematical Reviews and/or Referativny Zhurnal Matematika containing a review of the articles and books in the Literature list are given wherever possible. There are valuable reviews in Zentralblatt also.

The author wishes to express his gratitude to Prof. Leo F. Boron who translated and edited this book and to Prof. Dr. Larry E. Bobisud (University of Idaho) who read the typescript in great detail and made valuable suggestions for improvement. The expert assistance of Dr. William L. Voxman (University of Idaho and Escuela Politéchnica Nacional, Ecuador) and Dr. Kiyoshi Iséki (Kobe University) is also acknowledged.

Kishinyov
November 7, 1974

K. S. Sibirsky

FOREWORD TO THE SOVIET EDITION

The topological theory of dynamical systems (topological dynamics), whose foundations were laid by G. D. Birkhoff, developed especially intensively in the thirties and forties of the present century. Many of the results obtained during this time are contained in the monograph by V. V. Nemytsky and V. V. Stepanov [1] (the fourth chapter in the first edition and the fifth in the second) and in the book by W. H. Gottschalk and G. A. Hedlund [1]. In the fifties and sixties investigations were carried out which in particular allowed one to consider anew certain of the results obtained earlier.

In the present book we present the fundamentals of the theory of dynamical systems; however, the author strove as much as possible to also touch upon certain topics which are outside the scope of an 'introduction' to topological dynamics. In particular, in the last chapter various generalizations of the theory of dynamical systems are briefly considered.

The bibliography at the end of the book contains all the works cited in the body of this book; the bibliography published by W. H. Gottschalk [1] can very well serve as a supplementary source.

For a more complete acquaintance with the fundamental concepts of the theory of metric spaces which are used in this book we recommend the book by P. S. Aleksandrov [2]. We remark, however, that every time we deal with some concept or result from the above cited book we give the corresponding definitions and formulations of the results either in the body of the text itself or as a parenthetical remark. Therefore this book can be read by those who have not studied the theory of metric spaces.

The main content of this book has been presented in a lecture course which the author gave over several years at Kiev State University.

The author is deeply grateful to Viktor Vladimirovich Nemytsky (who perused the initial variant of the manuscript) for very useful advice as well as for the great amount of attention given to the development of research on the qualitative theory of differential equations in Kishinyov.

Valuable remarks by E. A. Barbashin and I. Yu. Bronshtein have been incorporated into the book. V. F. Rozhko and L. A. Chelysheva did a great amount of work preparing the manuscript for publication. The author is sincerely grateful to them all.

INTRODUCTION

The theory of differential equations originated at the end of the seventeenth century in the works of I. Newton, G. W. Leibniz and others.

During the first century of its existence, this theory consisted only of isolated methods of solving certain types of differential equations; but the problem of the existence of a solution and its representability in quadratures was posed already in the second. As a result of numerous investigations it became clear that integrability in quadratures is an extremely rare phenomenon and that the solution of many differential equations arising in applications cannot be expressed in quadratures. Also the methods of numerical integration of equations did not open the road to the general theory since these methods yield only one particular solution and this solution is obtained on a finite interval.

Applications – especially the problems of celestial mechanics – required the clarification of at least the nature of the behavior of integral curves in the entire domain of their existence without integration of the equation. In this connection, at the end of the last century there arose the qualitative theory of differential equations, the creators of which one must by all rights consider to be H. Poincaré and A. M. Lyapunov.

Poincaré [1, 2] formulated the problem of giving the most complete picture possible of the disposition of the integral curves of the equation $y' = f(x, y)$ or the system

$$\frac{dx}{dt} = P(x, y), \qquad \frac{dy}{dt} = Q(x, y) \tag{1}$$

in their entire domain of existence, starting from only the properties of the right members of the equations, without integrating them. He gave a classification of singular points, investigated the behavior of the integral curves in the neighborhoods of singular points, introduced the concept of

1

a limit cycle, and studied the course of integral curves on the surface of a torus. (A *limit cycle* is a closed integral curve which sufficiently close integral curves approach arbitrarily closely in a spiraling fashion. A *torus* is a surface obtained by rotating a circle about an axis which does not intersect it and lies in the plane of the circle.) These investigations are basically of a topological nature.

In 1901 I. Bendixson continued Poincaré's investigations on the basis of set-theoretic considerations. In particular, he discovered new types of singular points.

The investigations of A. M. Lyapunov concern the formulation and consideration of the general problem of the stability of a motion defined by a system of differential equations. In his famous doctoral dissertation, first published in 1892, Lyapunov [1] gave a rigorous definition of the concept of stability, pointed out cases when the problem of stability is solved by the first approximation, and also considered singular cases wherein it is impossible to judge stability on the basis of the first approximation.

Systems of the form (1) and of the more general form

$$\frac{dx_i}{dt} = \varphi_i(x_1, x_2, \ldots, x_n) \qquad (i = 1, 2, \ldots, n), \tag{2}$$

whose right members do not depend explicitly on the independent variable and are continuous in some region H of n-dimensional space R^n, were subsequently called *dynamical systems* by G. D. Birkhoff [1]. The variable t is usually called *time*, solutions $x_i = x_i(t)$ are called *motions*, and $x_1, x_2, \ldots, x_n$ are customarily called *coordinates* of the moving point in n-dimensional space R^n, which is called the *phase* space; the curves with parametric equations $\{x_i = x_i(t)\}$ are called *trajectories* of the motion.

[As a special case of the dynamical system (2) with $n = 1$ one can consider the differential equation of the form $dx/dt = \varphi(x)$. Here the phase space in which the point x moves is the line R^1.]

In the works of G. D. Birkhoff [3] (the first publication goes back to 1912), the theory of dynamical systems which Poincaré had in mind received a further extensive development. Birkhoff made use of topological methods to even a greater extent than Poincaré. Isolating such classes of motions as central and recurrent motions, Birkhoff actually laid the foundations of the theory of dynamical systems (topological dynamics). The basic problems of this theory were: (1) the study of solutions 'globally' (in the entire region of existence); (2) the study of solutions near singular points.

Many Soviet mathematicians made large contributions to the development of the topological theory of dynamical systems. Among them are A. A. Markov in whose work [1] there first appeared the general definition of a dynamical system, V. V. Stepanov, V. V. Nemytsky, G. F. Hilmy, M. V. Bebutov, A. G. Maier, E. A. Barbashin, I. Yu. Bronshtein, B. A. Shcherbakov, and others. Drawing upon modern mathematical methods of investigation, these scholars worked out a number of fundamental propositions in the topological theory of dynamical systems. New results in this area were also obtained by the American mathematicians M. Morse, G. A. Hedlund, W. H. Gottschalk, R. Ellis, J. Auslander, L. Auslander, and others.

Remark 1.   In contrast to arbitrary systems, in the theory of dynamical systems the right members of which do not depend explicitly on time (these are called *autonomous* systems), the following conditions are usually put on the functions $\varphi_i(x_1, x_2, \ldots, x_n)$ appearing in the system (2):

1. The uniqueness condition is satisfied in the entire region H, i.e., for any point $(x_{10}, x_{20}, \ldots, x_{n0}) \in H$ and any moment of time t_0 there exists a unique solution $x_i = x_i(t)$ of the system (2) which satisfies the initial conditions $x_i(t_0) = x_{i0}$ $(i = 1, 2, \ldots, n)$.

2. All the solutions in the region H exist in the infinite interval of time $-\infty < t < +\infty$.

The first condition is satisfied in particular if the functions $\varphi_i(x_1, x_2, \ldots, x_n)$ satisfy a Lipschitz condition in the region H. Regarding the second condition, it is not essential in studying the topological properties of trajectories since its satisfaction can always be attained by transition to the equivalent system

$$\frac{dx_i}{dt} = \varphi_i(x_1, x_2, \ldots, x_n)M(x_1, x_2, \ldots, x_n), \tag{3}$$

in which the trajectories are unaltered but the law of motion along them changes in view of the change in the velocity of motion (see, e.g., R. È. Vinograd [1], pages 100–101).

Remark 2.   The solutions of dynamical systems possess an important property which distinguishes them from the solutions of nonautomous systems (the right members of the latter depend explicitly on time).

Considering, for instance, the equations

$$\frac{dx}{dt} = x \quad \text{and} \quad \frac{dx}{dt} = 2tx$$

and denoting by $x = f(x_0, t)$ their particular solutions satisfying the initial condition $x = x_0$ at $t = 0$, we note that for the first equation $f(x_0, t) = x_0 e^t$ and

$$f(f(x_0, t_1), t_2) = f(x_0, t_1+t_2), \tag{4}$$

whereas for the second equation $f(x_0, t) = x_0 e^{t^2}$ and equality (4) does not hold.

All systems of the form (2) possess a property similar to (4). In fact, denoting the vector with components $x_1, x_2, \ldots, x_n$ by x and the vector-valued function with components $\varphi_1, \varphi_2, \ldots, \varphi_n$ by φ, we can write the system (2) in the form

$$\frac{dx}{dt} = \varphi(x), \tag{5}$$

and its solution satisfying the initial conditions $x_1 = x_{10}, x_2 = x_{20}, \ldots,$ $x_n = x_{n0}$ at $t = 0$ in the form $x = f(x_0, t)$ where $x_0 = \{x_{10}, x_{20}, \ldots, x_{n0}\}$ and f is some vector-valued function. In this case, the function $x = f(x_0, t-t_0)$ will obviously also satisfy equation (5) since its right member does not contain the variable t 'explicitly and consequently the substitution of $t-t_0$ for t is allowable in this equation. Fixing t_1 and setting $f(x_0, t_1) = x_1$, consider the two functions

$$x = f(x_1, t) \quad \text{and} \quad x = f(x_0, t_1+t),$$

which satisfy the equation (5) and the initial condition $x = x_1$ at $t = 0$. In view of the uniqueness, these functions must coincide for all t and in particular for $t = t_2$, i.e.,

$$f(f(x_0, t_1), t_2) = f(x_0, t_1+t_2),$$

which is what was required to be proved.

The properties listed above together with the condition that the solutions depend continuously on the coordinates of the initial point lie at the foundation of the general definition of a dynamical system.

Chapter I

GENERAL PROPERTIES OF DYNAMICAL SYSTEMS

§ 1   General definition of a dynamical system

Let X be a metric space (see, e.g., P. S. Aleksandrov [2], p. 227), i.e., X is an arbitrary set of elements (called points of the space X) and there is defined a nonnegative real-valued function $\rho(p, q)$ of the elements p and q in X, called the distance function and satisfying the following three conditions: (i) $\rho(p, q) = 0$ if and only if $p = q$ (identity axiom), (ii) $\rho(p, q) = \rho(q, p)$ (symmetry axiom), and (iii) $\rho(p, r) \leqq \rho(p, q) + \rho(q, r)$ for arbitrary $p, q, r \in X$ (triangle axiom).

Definition 1.1.   A *dynamical system* in the space X is a function $q = f(p, t)$ which assigns to each point p of the space X and to each real number t $(-\infty < t < +\infty)$ a definite point $q \in X$ and possesses the following three properties:

1.1.   Initial condition: $f(p, 0) = p$ for any point $p \in X$.

2.1.   Property of continuity in both arguments simultaneously:

$$\lim_{\substack{p \to p_0 \\ t \to t_0}} f(p, t) = f(p_0, t_0).$$

This property can be expressed in terms of sequences as follows: For an arbitrary sequence $\{p_n\}$ of points p_n in the space X which tends to p_0 and an arbitrary sequence $\{t_n\}$ of real numbers t_n which tends to t_0, the corresponding sequence $\{f(p_n, t_n)\}$ of points in the space X tends to $f(p_0, t_0)$.

In 'ε–δ' language this property means that for any $\varepsilon > 0$ there exists a $\delta > 0$ such that if $\rho(p, p_0) < \delta$ and $|t - t_0| < \delta$ then

$$\rho(f(p, t), f(p_0, t_0)) < \varepsilon.$$

3.1.   Group property:

$$f(f(p, t_1), t_2) = f(p, t_1 + t_2)$$

for any point $p \in X$ and any real t_1 and t_2.

We shall use the following notation:

$$R \equiv (-\infty, +\infty), R^+ \equiv [0, +\infty), R^- \equiv (-\infty, 0],$$

$$f(A, k) \equiv \bigcup_{p \in A, t \in K} f(p, t) \text{ for any } A \subseteq X, K \subseteq R,$$

$$\Sigma_A \equiv \overline{f(A, R)}, \Sigma_A^+ \equiv \overline{f(A, R^+)}, \Sigma_A^- \equiv \overline{f(A, R^-)}.$$

For every fixed value of the parameter t, which we shall call *time*, the function $q = f(p, t)$ defines a mapping of the space X into itself. Thus, a dynamical system defines a one-parameter family G of mappings of the space X into itself.

For fixed p, the function $f(p, t)$ is called a *motion*, and the set of points $q = f(p, t)$ for all t, $-\infty < t < +\infty$, i.e., the set $f(p, R)$, is called the *trajectory* of this motion.

The set of points $f(p, t)$ for fixed p and arbitrary $t \geqq 0$ ($t \leqq 0$), i.e., the set $f(p, R^+)(f(p, R^-))$, is called the *positive (negative) semitrajectory*, emanating from the point p.

The set of all points $f(p, t)$ for fixed p and arbitrary $t \in [T_1, T_2]$, i.e., the set $f(p, [T_1, T_2])$, is called a *segment* of the trajectory and the number $\tau = T_2 - T_1$ is called the *time interval* of this segment of the trajectory.

Example 1.1.   Consider the dynamical system defined by the system of differential equations

$$\frac{dx}{dt} = x, \quad \frac{dy}{dt} = y.$$

The solution of this system, satisfying the initial conditions $x = x_0$, $y = y_0$ at $t = 0$, is

$$x = x_0 e^t, \quad y = y_0 e^t.$$

Thus, if the point p has coordinates (x_0, y_0), then the point $q = f(p, t)$ will have coordinates $(x_0 e^t, y_0 e^t)$.

Clearly, the trajectory of the origin consists of a single point. But if $x_0 \neq 0$ or $y_0 \neq 0$, then as $t \to +\infty$ the point $q \to \infty$, and as $t \to -\infty$ the point $q \to 0$. Moreover, $x_0 y = y_0 x$. Thus, if the point p does not coincide with the origin then its trajectory is the half-axis emanating from the origin and passing through the point p (see Fig. 1).

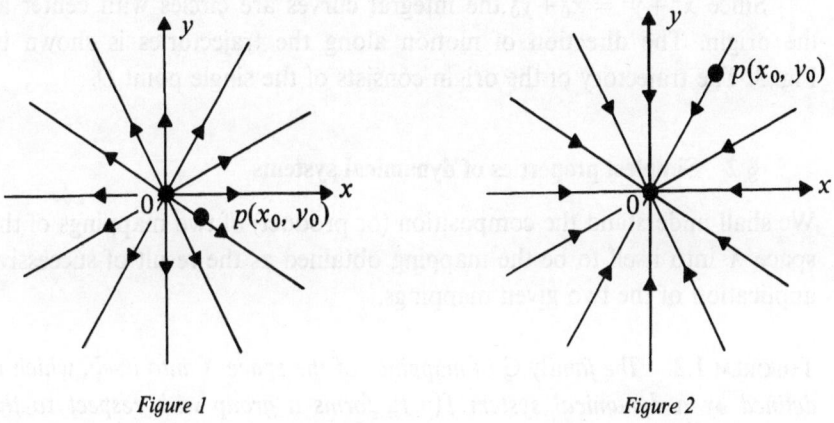

Figure 1   Figure 2

Example 2.1. $dx/dt = -x$, $dy/dt = -y$. This case can be reduced to that in the preceding example by replacing t by $-t$. Consequently, the trajectories of the motions will be the same as in Example 1.1 but the directions of the motions on them will be in the opposite direction. They will all enter the point $(0, 0)$ as $t \to +\infty$ (see Fig. 2).

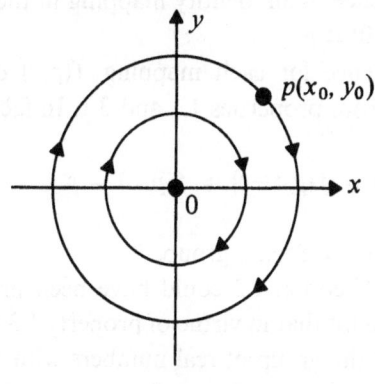

Figure 3

Example 3.1. $dx/dy = y$, $dy/dx = -x$. The solution of this system is

$$x = x_0 \cos t + y_0 \sin t,$$

$$y = -x_0 \sin t + y_0 \cos t.$$

Since $x^2 + y^2 = x_0^2 + y_0^2$, the integral curves are circles with center at the origin. The direction of motion along the trajectories is shown in Fig. 3. The trajectory of the origin consists of the single point O.

§ 2 Simplest properties of dynamical systems

We shall understand the composition (or product) of two mappings of the space X into itself to be the mapping obtained as the result of successive application of the two given mappings.

THEOREM 1.2. *The family G of mappings of the space X into itself, which is defined by a dynamical system $f(p, t)$, forms a group with respect to the operation of composition.*

Proof. (a) The composition of any two mappings $f(p, t_1)$ and $f(p, t_2)$ from the family G is, by virtue of property 3.1, a mapping $f(p, t_1 + t_2)$ which is again a mapping from the same family G (the commutativity of the composition of mappings in the family G follows from this same property);

(b) The operation of the composition of mappings from the family G is associative by virtue of the associativity of the addition of real numbers: $(t_1 + t_2) + t_3 = t_1 + (t_2 + t_3)$;

(c) The existence of an identity mapping in the family G follows from property 1.1: $f(p, 0) \equiv p$;

(d) The existence for each mapping $f(p, t)$ of an inverse mapping $f(p, -t)$ follows from properties 1.1 and 3.1. In fact,

$$f(f(p, t), -t) = f(p, t-t) = f(p, 0) = p.$$

Thus, the family G is indeed a group.

Remark 1.2. Theorem 1.2 could have been proved more simply had we taken into account that in virtue of property 1.3 the family G is a homomorphic image of the group of real numbers with respect to addition and that a homomorphic image of a group is a group.

THEOREM 2.2.   *The mappings in the group G are homeomorphisms of X onto itself.*

[A homeomorphism, or topological mapping, is any one-to-one onto bicontinuous mapping.]

In fact, the one-to-oneness of the mapping $q = f(p, t_0)$ of the space X onto itself is guaranteed by the existence of the inverse mapping $p = f(q, -t_0)$. The continuity of both of these mappings is guaranteed by property 2.1.

THEOREM 3.2.   *The following relations hold for arbitrary $A_\alpha \subseteq X$, $K_\beta \subseteq R$:*

$$f(\bigcup_\alpha A_\alpha, \bigcup_\beta K_\beta) = \bigcup_{\alpha,\beta} f(A_\alpha, K_\beta), \tag{1.2}$$

$$f(\bigcap_\alpha A_\alpha, t) = \bigcap_\alpha f(A_\alpha, t), \tag{2.2}$$

$$f(p, \bigcap_\beta K_\beta) \subseteq \bigcap_\beta f(p, K_\beta), \tag{3.2}$$

$$f(f(p, K_1), K_2) = f(p, K_1 + K_2), \tag{4.2}$$

where

$$K_1 + K_2 \equiv \bigcup_{t_1 \in K_1, t_2 \in K_2} (t_1 + t_2).$$

We shall prove, for example, relation (2.2). Let

$$q \in f(\bigcap_\alpha A_\alpha, t). \tag{5.2}$$

Then

$$f(q, -t) \in f(f(\bigcap_\alpha A_\alpha, t), -t) = f(\bigcap_\alpha A_\alpha, 0) = \bigcap_\alpha A_\alpha \subseteq A_\alpha$$

for arbitrary α. But then, for arbitrary α,

$$q = f(f(q, -t), t) \in f(A_\alpha, t),$$

and consequently

$$q \in \bigcap_\alpha f(A_\alpha, t). \tag{6.2}$$

Conversely, suppose (6.2) is satisfied. Then, for any α, $q \in f(A_\alpha, t)$ and $f(q, -t) \in A_\alpha$. Consequently,

$$f(q, -t) \in \bigcap_\alpha A_\alpha,$$

from which (5.2) follows.

That $\subseteq$ in relation (3.2) cannot be replaced by equality follows from the following example.

Consider the uniform periodic motion $q = f(p, t)$ on the circle L with period 1: $f(p, t+1) = f(p, t)$ for all $t \in R$.

Let $K_1 = [0, 1]$, $K_2 = [1, 2]$. Then $K_1 \cap K_2 = \{1\}$ and $f(p, K_1 \cap K_2) = f(p, 1) = p$, whereas

$$f(p, K_1) \cap f(p, K_2) = f(p, K_1) = f(p, K_2) = f(p, R) = L.$$

We shall now establish two lemmas on the continuity of distance.

LEMMA 1.2. *If* $\lim_{n \to \infty} q_n = q_0$ *and* $\lim_{n \to \infty} p_n = p_0$, *then*

$$\lim_{n \to \infty} \rho(q_n, p_n) = \rho(q_0, p_0), \tag{7.2}$$

i.e., passage to the limit commutes with distance.

Proof. In fact, on the basis of properties of distance

$$\rho(q_n, p_n) \leq \rho(q_n, q_0) + \rho(q_0, p_n) \tag{8.2}$$

$$\leq \rho(q_n, q_0) + \rho(q_0, p_0) + \rho(p_0, p_n).$$

Interchanging p_0 with p_n, q_0 with q_n in (8.2), we find that

$$\rho(q_n, p_n) \geq \rho(q_0, p_0) - \rho(q_n, q_0) - \rho(p_0, p_n). \tag{9.2}$$

Taking into consideration that $\lim_{n\to\infty} \rho(q_n, q_0) = 0$ and $\lim_{n\to\infty} \rho(p_0, p_n) = 0$, we obtain (7.2) from (8.2) and (9.2).

We have thus proved that distance is continuous.

LEMMA 2.2. *If* $\lim\limits_{n\to\infty} p_n = p_0$ *and* $A \subseteq X$, *then*

$$\lim_{n\to\infty} \rho(p_n, A) = \rho(p_0, A).$$

[The distance between the point r and the set A is understood to be $\inf_{p\in A} \rho(r, p)$.]

Proof. In fact, since, for any point $p \in A$,

$$\rho(p_n, p) \leq \rho(p_n, p_0) + \rho(p_0, p),$$

then also

$$\rho(p_n, A) \leq \rho(p_n, p_0) + \rho(p_0, p),$$

and then

$$\rho(p_n, A) \leq \rho(p_n, p_0) + \rho(p_0, A). \tag{10.2}$$

Interchanging the roles of p_n and p_0, we obtain

$$\rho(p_n, A) \geq \rho(p_0, A) - \rho(p_n, p_0). \tag{11.2}$$

Since $\rho(p_n, p_0) \to 0$ as $n \to \infty$, from (10.2) and (11.2) we obtain that

$$\rho(p_n, A) \to \rho(p_0, A).$$

THEOREM 4.2 (INTEGRAL CONTINUITY CONDITION). *For any point* $p \in X$, *any number* $T > 0$ *and any* $\varepsilon > 0$, *there exists a* $\delta > 0$ *such that* $\rho(f(q, t), f(p, t)) < \varepsilon$ *for all* $q \in X$ *and* $t \in R$ *which satisfy the inequalities* $\rho(q, p) < \delta$ *and* $0 \leq t \leq T(-T \leq t \leq 0)$.

[In other words, if sufficiently close initial points are chosen then during a prescribed interval of time T the distance between the simultaneous positions of moving points will remain less than a given positive number ε.]

Proof. We shall prove this property by assuming the contrary and arriving at a contradiction. Assume that for some point $p_0 \in X$, some number $T_0 > 0$ and some $\varepsilon_0 > 0$ there does not exist a corresponding $\delta > 0$ which satisfies the condition of the theorem. Take a sequence $\{\delta_n\}$ of positive numbers which converges to zero. According to our assumption, for every natural number n there exists at least one point q_n such that $\rho(q_n, p_0) < \delta_n$ and at least one number $t_n \in [0, T_0]$ for which

$$\rho(f(q_n, t_n), f(p_0, t_n)) \geq \varepsilon_0. \tag{12.2}$$

From the fact that $\rho(q_n, p_0) < \delta_n \to 0$ as $n \to \infty$ it follows that $\{q_n\} \to p_0$.

The sequence $\{t_n\}$ of time values is bounded. Hence by the Bolzano-Weierstrass theorem it contains a convergent subsequence. In order not to complicate the notation, assume that the sequence $\{t_n\}$ itself converges. Let $\lim_{n \to \infty} t_n = t_0$. Then, by virtue of the joint continuity of the function $f(p, t)$,

$$\lim_{n \to \infty} f(q_n, t_n) = f(p_0, t_0) \quad \text{and} \quad \lim_{n \to \infty} f(p_0, t_n) = f(p_0, t_0).$$

Passing to the limit, on the basis of Lemma 1.2, in inequality (12.2) yields $0 \geq \varepsilon_0$. The contradiction thus obtained proves the theorem.

§ 3 The classification of motions and trajectories

Consider an arbitrary motion $f(p, t)$. Three cases are possible:

1. $f(p, t) \equiv p$ (for all $t \in R$). Such a motion is called a *rest* motion. In this case the point p is called a *rest* (*critical, equilibrium* or *stationary*) point.

2. $f(p, t) \not\equiv p$, but there exist numbers $t_1, t_2, t_1 \neq t_2$, such that $f(p, t_2) = f(p, t_1)$. Assuming for definiteness that $t_2 > t_1$, let $t_2 - t_1 = \tau$ ($\tau > 0$). Then

$$f(p, t+\tau) = f(p, t+t_2-t_1) = f(f(p, t_2), t-t_1)$$

$$= f(f(p, t_1), t-t_1) = f(p, t_1+t-t_1) = f(p, t)$$

for any $t \in R$. In this case the motion is said to be *periodic* and the number τ is a *period* of the motion.

Clearly, along with τ the motion considered also allows periods of the form $n\tau$ $(n = 0, \pm 1, \pm 2, \ldots)$. We shall show that among all positive periods of the motion $f(p, t)$ there exists a smallest one. To this end, consider the set $\{\tau\}$ of all positive periods and let $\inf\{\tau\} = \tau_0$. Then there exists a sequence of positive periods $\{\tau_n\} \to \tau_0$. Passing to the limit as $n \to \infty$ in the equality $f(p, t+\tau_n) = f(p, t)$, we obtain that $f(p, t+\tau_0) = f(p, t)$ for all $t \in R$, i.e., τ_0 also is a period of the motion $f(p, t)$. It remains to prove that $\tau_0 \neq 0$. Suppose the contrary, i.e., that $\{\tau_n\} \to 0$.

Consider the point p and an arbitrary neighborhood $S(p, \varepsilon)$ of p. [$S(p, \varepsilon)$ denotes the set $\{q \in X | \rho(q, p) < \varepsilon\}$ and $\bar{S}(p, \varepsilon)$ denotes its closure.] By virtue of the continuity of the function $f(p, t)$ with respect to t, for given $\varepsilon > 0$ there exists a $\delta > 0$ such that

$$\rho(f(p, t), p) < \varepsilon \qquad (1.3)$$

for all $|t| < \delta$. Clearly, there exists a $\tau_n < \delta$. Then the inequality (1.3) is satisfied for all $t \in [0, \tau_n]$ and, since τ_n is a period, it must be satisfied for all $t \in R$. Thus, $f(p, t) \in S(p, \varepsilon)$ for all $t \in R$. From this, since ε is arbitrary, we obtain that $f(p, t) = p$ for all t, and this contradicts the condition that $f(p, t) \not\equiv p$.

In the sequel we shall understand the *period* of a motion $f(p, t)$ to be the smallest positive period.

3. $f(p, t_2) \neq f(p, t_1)$ for $t_2 \neq t_1$. Such a motion is called *nonsingular* in contrast to motions of the first two types which are further called *singular* motions.

Thus there exist three distinct types of motions and three corresponding distinct types of trajectories:

1. In the case of a rest motion the trajectory is a point.

2. In the case of a periodic motion with period τ the trajectory is a simple closed curve, i.e., a 1–1 bicontinuous image of the segment $[0, \tau]$ on the real line whose endpoints have been identified. In other words, one can say that the trajectory of a periodic motion is homeomorphic to a circle. [See P. S. Aleksandrov [2], page 321, Theorem 11: A 1–1 bicontinuous mapping of a compactum is a homeomorphism. A set $A \subseteq X$ is said to be sequentially compact if any infinite sequence of points in the set A contains a subsequence which converges to some point of the space X. In R^n sequential compactness is equivalent to boundedness. A closed sequentially compact set is said to be a compact set or a compactum.]

3. In the case of a nonsingular motion the trajectory is a 1–1 bicontinuous image of the real line.

The subsequent four theorems express certain properties of arcs and trajectories of a dynamical system.

THEOREM 1.3.   *An arbitrary finite arc of a trajectory is a compact set.*

Proof. Consider the arc $f(p, [T_1, T_2])$ and an arbitrary sequence of points $\{q_n\}$ on it. Since $q_n \in f(p, [T_1, T_2])$, there exists a real number $t_n \in [T_1, T_2]$ such that $q_n = f(p, t_n)$. Since it is bounded, the sequence $\{t_n\}$ has, by the Bolzano-Weierstrass theorem, a convergent subsequence $\{t_{n_k}\}$. Let $\lim_{k \to \infty} t_{n_k} = t_0$. Then $t_0 \in [T_1, T_2]$ and therefore

$$f(p, t_0) \in f(p, [T_1, T_2]).$$

If we now consider the corresponding subsequence of points $\{q_{n_k}\}$, then, by virtue of the continuity of the function $f(p, t)$,

$$\lim_{k \to \infty} q_{n_k} = \lim_{k \to \infty} f(p, t_{n_k}) = f(p, t_0) \in f(p, [T_1, T_2]).$$

This has simultaneously proved that the set $f(p, [T_1, T_2])$ is both sequentially compact and closed, i.e., that it is compact. [This theorem also follows from the fact that the arc $f(p, [T_1, T_2])$ can be considered as a continuous image of the compact set of numbers $[T_1, T_2]$ (see P.S. Aleksandrov [2], page 320, Theorem 8). Moreover, in the case of a nonsingular motion the arc $f(p, [T_1, T_2])$, being a 1–1 bicontinuous image of the compactum $[T_1, T_2]$, is homeomorphic to the segment $[T_1, T_2]$ (loc. cit., Theorem 11, page 321).]

THEOREM 2.3.   *If $q \in f(p, R)$, then $f(q, R) = f(p, R)$.*

Proof. Let $q \in f(p, R)$. Then, taking (4.2) into account,

$$f(q, R) \subseteq f(f(p, R), R) = f(p, R+R) = f(p, R).$$

On the other hand, the point q, being a point on the trajectory of the point p, is representable in the form $q = f(p, t_0)$. Then

$$p = f(q, -t_0) \in f(q, R)$$

and by what has been proved $f(p, R) \subseteq f(q, R)$.

The inclusions just obtained show that $f(p, R) = f(q, R)$.

THEOREM 3.3.   *Distinct trajectories of a dynamical system do not intersect.*

Proof. To prove this we shall show that if two trajectories $f(p, R)$ and $f(q, R)$ have a point r in common, then they coincide. In fact, if $r \in f(p, R)$ and $r \in f(q, R)$, then, on the basis of Theorem 2.3,

$$f(r, R) = f(p, R) \quad \text{and} \quad f(r, R) = f(q, R),$$

and consequently $f(p, R) = f(q, R)$.

THEOREM 4.3.   *The motion of a point p uniquely defines the motion of all points on the trajectory $f(p, R)$.*

Proof. In fact, if the motion $f(p, t)$ of the point p is known then the motion of an arbitrary point $q = f(p, t_0) \in f(p, R)$ is determined as follows:

$$f(q, t) = f(f(p, t_0), t) = f(p, t_0 + t).$$

§ 4   Invariant sets

Definition 1.4.   A set A of points in the space X is *invariant with respect to a given dynamical system $f(p, t)$*, or simply *invariant*, if it is mapped into itself by all mappings of the given system, i.e., if $f(A, t) \subseteq A$ for all $t \in R$. Furthermore, also $f(A, -t) \subseteq A$. Applying to the latter inclusion a mapping with parameter t, we find that $A \subseteq f(A, t)$, and hence $f(A, t) = A$.

The entire trajectory can serve as an example of an invariant set since, on the basis of (4.2),

$$f(f(p, R), t) = f(p, R+t) = f(p, R).$$

THEOREM 1.4.   *The union of an arbitrary collection of invariant sets is invariant.*

Proof. Let $A = \bigcup_\alpha A_\alpha$, where the A_α are invariant sets, i.e., $f(A_\alpha, t) = A_\alpha$ for all $t \in R$. Then, taking (1.2) into account,

$$f(A, t) = f(\bigcup_\alpha A_\alpha, t) = \bigcup_\alpha f(A_\alpha, t) = \bigcup_\alpha A_\alpha = A$$

and so the set A is invariant.

From Theorem 1.4 it follows in particular that any collection of entire trajectories is an invariant set. One can also prove the converse, i.e., every invariant set consists of entire trajectories. In fact, let $p \in A$, where A is an invariant set. Then, for any $t \in R$ whatever,

$$f(p, t) \in f(A, t) = A.$$

It follows that $f(p, R) \subseteq A$.

We have thus proved the next theorem.

THEOREM 2.4   (CHARACTERIZING PROPERTY OF AN INVARIANT SET). *A necessary and sufficient condition for a set to be invariant is that it consist of entire trajectories, i.e., that this set contains together with each of its points also the trajectory through the point.*

THEOREM 3.4.   *The nonempty intersection of an arbitrary collection of invariant sets is invariant.*

   Proof. Let $A = \bigcap_\alpha A_\alpha$, where the A_α are invariant sets. Then, taking (2.2) into account,

$$f(A, t) = f(\bigcap_\alpha A_\alpha, t) = \bigcap_\alpha f(A_\alpha, t) = \bigcap_\alpha A_\alpha = A,$$

and so the set A is invariant.

THEOREM 4.4.   *The complement $X \setminus A$ of an invariant set $A \subset X$ is invariant.*

   Proof. Let $A \subset X$ and $f(A, t) = A$ for every $t \in R$. Then for any point $q \in X \setminus A$ and arbitrary $t \in R$ we have that $f(q, t) \in X \setminus A$.

   In fact, if we had $f(q, t) \in A$, then $q \in f(A, -t) = A$, which is impossible since $q \in X \setminus A$. From the fact that $f(q, t) \in X \setminus A$ for any $t \in R$ it follows that $f(q, R) \in X \setminus A$. Thus, $X \setminus A$ consists of entire trajectories and by Theorem 2.4 it is invariant.

THEOREM 5.4.   *The closure of any invariant set is invariant.*

   Proof. Let A be an invariant set. We shall show that its closure $\bar{A}$ consists of entire trajectories. Let $p \in \bar{A}$. Thus there exists a sequence of points in A (among which there can be equal terms) $\{q_n\} \to p$. Since A is invariant we have that $f(q_n, t) \in A$ for arbitrary $t \in R$. But $\{f(q_n, t)\} \to f(p, t)$. Hence $f(p, t) \in \bar{A}$ and, since t is arbitrary, we have that $f(p, R) \subseteq \bar{A}$ and this means that $\bar{A}$ is invariant (see Theorem 2.4).

Remark 1.4.   Clearly, every dynamical system $f(p, t)$ defined in a space X uniquely defines a dynamical system on an arbitrary invariant subset $A \subseteq X$.

Definition 2.4.   A set A is *positively* (*negatively*) *invariant* if $f(A, t) \subseteq A$ for arbitrary $t \in R^+$ ($t \in R^-$).

It is not difficult to show that for positively (negatively) invariants sets theorems analogous to Theorems 1.4, 2.4, 3.4, 5.4 hold and that the complement of a positively invariant set is negatively invariant.

§ 5   Theorems on rest points

THEOREM 1.5.   *The set of rest points is closed.*

Proof.   Consider the sequence of rest points $\{p_n\}$ which converges to the point p_0. The equality $f(p_n, t) = p_n$ is valid for any n and t; passing to the limit in this equality as $n \to \infty$ we obtain that $f(p_0, t) = p_0$ for all $t \in R$, i.e., p_0 is also a rest point.

THEOREM 2.5.   *No motion can enter a rest point for a finite value of the time.*

Proof.   Let p be a rest point and assume that some motion $f(q, t)$ ($q \neq p$) entered the rest point p at some value of time $t = t_0$, i.e., $f(q, t_0) = p$. Then, on the basis of Theorem 2.3, $f(q, R) = f(p, R) = p$, which contradicts the fact that $q \neq p$. This theorem is also a direct consequence of Theorem 3.3.

THEOREM 3.5.   *If for some number $T > 0$ and for any neighborhood of the point p there exists a segment of the trajectory of time length T which is entirely contained in this neighborhood,* then p is a rest point.

Proof.   We shall prove this theorem by assuming the opposite and arriving at a contradiction. Suppose that p is not a rest point. Then there exists a t_0 on $[0, T]$ such that $f(p, t_0) \neq p$. Let $f(p, t_0) = p_0$ and $\rho(p, p_0) = d$ ($d > 0$). Corresponding to the number $d/2$, there exists by the continuity of the function $f(p, t)$ a $\delta > 0$ such that

$$\rho(f(q, t_0), f(p, t_0)) < \frac{d}{2} \tag{1.5}$$

for any point $q \in S(p, \delta)$. Furthermore we can always assume that $\delta < d/2$. Inequality (1.5) can be rewritten in the form

$$\rho(f(q, t_0), p_0) < \frac{d}{2}.$$ (2.5)

Using the triangle inequality and inequality (2.5), we obtain that

$$d = \rho(p, p_0) \leqq \rho(p, f(q, t_0)) + \rho(f(q, t_0), p_0) < \rho(p, f(q, t_0)) + \frac{d}{2}.$$

Then $\rho(p, f(q, t_0)) > d/2$. Thus, for any point $q \in S(p, \delta)$ we have that $f(q, t_0) \notin S(p, \delta)$. Since $0 \leqq t_0 \leqq T$, $S(p, \delta)$ contains no segment of a trajectory of time length T, contradicting the condition in the theorem. Hence p is a rest point.

COROLLARY 1.5. *If either* $\lim_{t \to +\infty} f(q, t) = p$ *or* $\lim_{t \to -\infty} f(q, t) = p$ *exists, then p is a rest point.*
    Proof. If $\lim_{t \to +\infty} f(q, t) = p$ $(\lim_{t \to -\infty} f(q, t) = p)$, then for any $\varepsilon > 0$ there exists a moment of time t_0 such that every semitrajectory $f(f(q, t_0), R^+)$ $(f(f(q, t_0), R^-))$ lies in the ε-neighborhood of the point p. On the basis of Theorem 3.5, the point p is a rest point. The proof is similar for the case $\lim_{t \to -\infty} f(q, t) = p$.

THEOREM 4.5. *If every neighborhood of the point p contains points with arbitrarily small periods, then p is a rest point.*
    Proof. Let $\{p_n\} \to p$, $f(p_n, \tau_n) = p_n$, $\{\tau_n\} \to 0$ $(\tau_n > 0)$ and $\varepsilon > 0$. Since the function $f(p, t)$ is continuous, there exists a $T > 0$ such that

$$\rho(f(p, t), f(p, 0)) < \frac{\varepsilon}{2}$$ (3.5)

for all $t \in [0, T]$.
    Corresponding to $\varepsilon/2$, the point p and the number T, we choose a $\delta > 0$ guaranteed by the integral continuity condition. It can always be assumed that $\delta < \varepsilon/2$. According to our assumption, there exists a natural number n such that $p_n \in S(p, \delta)$ and $\tau_n < T$.
    Then according to the choice of δ, we will have, for all $t \in [0, \tau_n]$, that

$$\rho(f(p_n, t), f(p, t)) < \frac{\varepsilon}{2}. \qquad (4.5)$$

It follows from (4.5) and (3.5) that, for all $t \in [0, \tau_n]$,

$$\rho(f(p_n, t), p) < \varepsilon.$$

Since according to our assumption τ_n is a period of the motion $f(p_n, t)$, we have that

$$f(p_n, R) \subseteq S(p, \varepsilon).$$

By Theorem 3.5 the point p is a rest point.

THEOREM 5.5. *Every positively invariant set M which is homeomorphic to a closed ball in the Euclidean space R^n contains a rest point.*

Proof. Take a sequence of positive numbers $\{t_k\}$ which converges to zero. Since M is positively invariant, for every k the mapping $f(p, t_k)$ maps the set M into itself. According to Brouwer's fixed point theorem this mapping has at least one fixed point, i.e., there exists a point $p_k \in M$ such that $f(p_k, t_k) = p_k$. [See P. S. Aleksandrov [1], page 584, Theorem [5:4]: Every continuous mapping of an n-dimensional cell into itself has at least one fixed point (an n-dimensional cell is a set which is homeomorphic to a closed ball in Euclidean space R^n).] Since an n-dimensional cell is compact, we can consider the sequence $\{p_k\}$ to be convergent. Let $\{p_k\} \to p$. Clearly, $p \in M$. By Theorem 4.5, the point p is a rest point. This completes the proof of the theorem.

It is well known that if L is a simple closed curve in the plane R^2 and G is the region bounded by L, then $\bar{G}$ is homeomorphic to a disc, i.e., it is a two-dimensional cell. [Compare, e.g., Riemann's theorem on conformal mapping and the theorem on correspondence of boundaries (I. I. Privalov [1], pages 385 and 392).] Furthermore, Theorem 5.5 implies the following corollary.

COROLLARY 2.5. *In the plane, in the interior of a region bounded by a trajectory of a periodic motion, there exists at least one rest point.*

Theorem 5.5 was proved in the work of N. P. Bhatia, A. C. Lazer and G. P. Szegö [1, 2]. The content of Corollary 2.5 for systems of differential equations is usually called the Poincaré-Bendixson theorem.

19

THEOREM 6.5. *Every dynamical system on the sphere S^{2k} possesses at least one rest point.*

Proof. Consider a sequence of positive numbers $\{t_n\}$ which converges to zero and the corresponding sequence of homeomorphisms of the sphere S^{2k} onto itself (see Theorem 2.2):

$$\{f(p, t_n)\}. \tag{5.5}$$

According to the theorem to the effect that for every continuous mapping of the sphere S^{2k} into itself there exists at least one fixed point or a point which is mapped into its diametrically opposite point, for every n there exists on S^{2k} at least one point p_n such that

$$f(p_n, t_n) = p_n \tag{6.5}$$

or

$$f(p_n, t_n) = p'_n, \tag{7.5}$$

where p'_n is the point diametrically opposite p_n. [See P. S. Aleksandrov [1], page 584, Theorem [5 : 52]. We understand the sphere S^{2k} to be the set of points in R^{2k+1} whose coordinates satisfy the equation $x_1^2 + x_2^2 + \ldots + x_{2k+1}^2 = r^2$.] Since S^{2k} is compact, it can be assumed that the sequence $\{p_n\}$ converges. Let

$$\lim_{n \to \infty} p_n = p_0.$$

We shall prove that p_0 is a rest point. Indeed, it follows from the fact that $\{t_n\} \to 0$ and $\{p_n\} \to p_0$ that

$$\lim_{n \to \infty} f(p_n, t_n) = f(p_0, 0) = p_0.$$

In this connection, equalities (7.5) can hold only for a finite number of values of n. Otherwise in the passage to the limit in (7.5) in corresponding sub-sequences we would obtain that $p_0 = p'_0$ where p'_0 is the point diametrically opposite to p_0. Without loss of generality, we may assume that equalities (6.5) are satisfied for all n. Then, according to Theorem 4.5, the point p_0 is a rest point. This completes the proof of the theorem.

Whether or not there exists a dynamical system without rest points and periodic trajectories on the sphere S^3 is at present an open question. [See, e.g., S. Smale [1].]

§ 6   Dynamical systems on the real line. The isomorphism of dynamical systems

Example 1.6. $dx/dt = x$; $x = x_0 e^t$. The dynamical system thus defined on the real x-axis contains only three trajectories: a rest point O and two infinite intervals $(-\infty, 0)$ and $(0, +\infty)$. The motion along the first of these occurs in the negative direction and along the second it occurs in the positive direction (see Fig. 4).

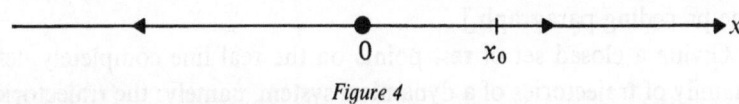

Figure 4

Example 2.6. $dx/dt = -x$; $x = x_0 e^{-t}$. The trajectories of the motion are the same as in the preceding example, but the motion along them is in the opposite direction (see Fig. 5).

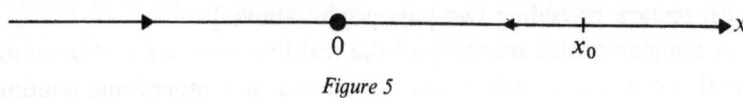

Figure 5

Example 3.6. $dx/dt = |x|$. Here the trajectories are the same as in the preceding examples. The motion along them is shown in Fig. 6.

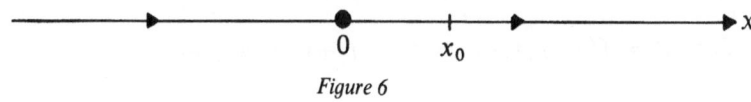

Figure 6

Example 4.6. $dx/dt = 0$; $x = x_0$. All of the points on the real line are rest points.

Example 5.6. $dx/dt = 1$; $x = x_0 + t$. In this case there is only one trajectory which coincides with the entire real line. The motion on it is uniform from left to right.

It is easy to see that it is impossible to map a circle homeomorphically onto the real line or onto any subset of the real line. For, suppose the opposite and take two distinct points A and B on the circle and denote their images on the real line under some homeomorphism by a and b respectively. The points A and B divide the circle into two arcs ACB and ADB which are homeomorphic to segments. Furthermore the image of each of these arcs must coincide with the segment $[a, b]$ of the real line and hence the one-oneness of the mapping is disturbed. [See N. Bourbaki [1], chapter IV, §2, page 22: A 1–1 bicontinuous mapping of an interval in R^1 into R^1 is strictly monotonic and is a homeomorphism of the interval onto its image; the latter is an interval in R^1.]

In view of the fact just established, periodic motions cannot exist. In this case all motions are divided into rest motions and nonsingular motions, and trajectories are either points or intervals. [See the parenthetical remark in the preceding paragraph.]

Giving a closed set of rest points on the real line completely defines the family of trajectories of a dynamical system, namely: the trajectories of nonsingular motions coincide with the intervals adjacent to the closed set of rest points.

For a complete definition of a dynamical system on the real line it is obviously sufficient to define a motion on each of the intervals, i.e., to homeomorphically map the real line onto each of them. [See the parenthetical remark preceding two paragraphs, above.]

A homeomorphic mapping of the real line $-\infty < t < +\infty$ onto the interval $a < x < b$ is defined only by means of a monotonic continuous function $x = \psi(t)$ such that $x \to b$ (or $x \to a$) as $t \to +\infty$ and $x \to a$ (or $x \to b$) as $t \to -\infty$. [See the parenthetical remark preceding three paragraphs, above.] Let $x_0 = \psi(0)$. Set $f(x_0, t) = \psi(t)$. Then the motion $f(x_1, t)$ of the point x_1 $(a < x_1 = \psi(t_1) < b)$ is defined as follows (see Theorem 4.3):

$$f(x_1, t) = f(f(x_0, t_1), t) = f(x_0, t_1 + t) = \psi(t_1 + t).$$

Clearly, not every dynamical system on the real line can be defined by means of a differential equation $dx/dt = \varphi(x)$ since there exist continuous monotonic functions $x = \psi(t)$ which do not have a derivative at all points of the interval $-\infty < t < +\infty$.

Definition 1.6. Two dynamical systems $g(p, t)$ and $h(x, t)$ $(p \in X_1, x \in X_2, t \in R)$ are said to be *isomorphic* or *topologically equivalent* if there exists a homeomorphism $x = \xi(p)$ of the space X_1 onto the space X_2 such that

$$\xi(g(p, t)) = h(\xi(p), t) \tag{1.6}$$

for all $p \in X_1$ and $t \in R$.

If the mapping ξ is only continuous, then when condition (1.6) is satisfied the system $g(p, t)$ is said to be *homomorphic* to the system $h(x, t)$.

The simplest example of a homomorphism is the situation when the space X_2 consists of one rest point.

Remark 1.6.   In the definition of an isomorphism the two dynamical systems enjoy equal rights since manifestly if ξ is a homeomorphism then condition (1.6) is equivalent to the satisfaction of the equality

$$\xi^{-1}(h(x, t)) = g(\xi^{-1}(x), t)$$

for all $x \in X_2$ and $t \in R$.

LEMMA 1.6.   *Under a homomorphism of dynamical systems the image of the segment $g(p, [t_1, t_2])$ of the trajectory of the point $p \in X_1$ is the segment $h(x, [t_1, t_2])$ of the trajectory of the point $x = \xi(p) \in X_2$.*

Proof. Let $p \in X_1$ and $x = \xi(p)$. Then equality (1.6) shows that for any $t \in [t_1, t_2]$ the image of the point $g(p, t)$ belongs to the segment $h(x, [t_1, t_2])$. On the other hand, every point $h(x, t)$ of the latter segment is the image of the point $g(p, t)$ on the segment $g(p, [t_1, t_2])$ under the mapping ξ.

The following lemma is proven analogously.

LEMMA 2.6.   *A homomorphism of dynamical systems maps a trajectory (positive semitrajectory) into a trajectory (positive semitrajectory)*

This implies the following corollary.

COROLLARY 1.6.   *A homomorphism of dynamical systems maps a rest point into a rest point.*

Taking Lemmas 1.6 and 2.6 and Remark 1.6 into consideration, it is easy to establish the next corollary.

COROLLARY 2.6.   *A homomorphism of dynamical systems maps a periodic point into a periodic point and furthermore in the case of an isomorphism the period is retained.*

Using Lemma 2.6 and Theorem 2.4 we can now prove the following corollary.

COROLLARY 3.6.   *A homomorphism of dynamical systems maps an invariant set into an invariant set.*

Finally, using Corollary 1.6 and taking Remark 1.6 into account we can easily obtain the next corollary.

COROLLARY 4.6   *A necessary condition for two dynamical systems $g(p, t)$ and $h(x, t)$ with sets of rest points M_1 and M_2 respectively to be isomorphic is that either one of the following two conditions be satisfied:*
  (1) $M_1 = M_2 = \phi$;
  (2) $M_1 \neq \phi, M_2 \neq \phi$ *and there exists a homeomorphism of X_1 onto X_2 which maps M_1 onto M_2.*

The following lemma holds.

LEMMA 3.6.   *Two periodic dynamical systems defined on circles are isomorphic if and only if their motions have the same period.*

Proof. The proof of necessity is obtained by Corollary 2.6. For the proof of sufficiency, on the other hand, if the periods are the same then taking any two points q and x on the circles X_1 and X_2 we set

$$\xi(g(q, t)) = h(x, t)$$

for any $t \in R$. It is easy to see that ξ will be a homeomorphism of the circle X_1 onto X_2 satisfying condition (1.6). We shall prove the latter. Setting $p = g(q, t_0)$, we have that

$$\xi(g(p, t)) = \xi(g(q, t+t_0)) = h(x, t+t_0)$$
$$= h(h(x, t_0), t) = h(\xi(g(q, t_0)), t) = h(\xi(p), t).$$

LEMMA 4.6.   *Any two dynamical systems without rest points defined on the real line are isomorphic.*

Proof. Let $g(p, t)$ and $h(x, t)$ be two dynamical systems defined on R^1. For arbitrary $t \in R$, set

$$\xi(g(0, t)) = h(0, t).$$

It is easy to see that ξ is a homeomorphism of R^1 onto R^1. Furthermore, if $p = g(0, t_0)$, then

$$\xi(g(p, t)) = \xi(g(0, t+t_0)) = h(0, t+t_0)$$

$$= h(h(0, t_0), t) = h(\xi(g(0, t_0)), t)$$

$$= h(\xi(p), t),$$

i.e., condition (1.6) is satisfied.

Example 6.6. The dynamical system defined on the real line R^1 by the differential equation

$$\frac{dx}{dt} = 1 + |x|$$

does not have rest points and consequently it is isomorphic to the dynamical system of Example 5.6. The motions in this example are such that

$$\int_{x_0}^{x} \frac{d\alpha}{1 + |\alpha|} = t.$$

It follows in particular that for $x_0 > 0$ and $t > 0$,

$$x = (1 + x_0)e^t - 1. \tag{2.6}$$

THEOREM 1.6. *A necessary and sufficient condition for the two dynamical systems $g(p, t)$ and $h(x, t)$ defined in R^1 to be isomorphic is that there exist a homeomorphic mapping $x = \eta(p)$ of the real line R^1 onto R^1 which carries the sets of rest points M_1 and M_2 of these systems one into the other so that the direction of motion of corresponding intervals adjacent to M_1 and M_2 be compatible (i.e., coincide if $\eta(p)$ increases, and be oppositely directed if $\eta(p)$ decreases).*

[Recall that a homeomorphic mapping of R^1 onto R^1 is monotonic.]

Proof. Suppose the systems $g(p, t)$ and $h(x, t)$ defined in R^1 are iso-morphic. Then, by virtue of (1.6), the homeomorphism $x = \xi(p)$ carries the sets of rest points M_1 and M_2 of these systems one into the other. In view of the fact that a homeomorphism of R^1 onto R^1 must be monotonic, it follows from (1.6) that the direction of motion in the corresponding intervals

25

adjacent to M_1 and M_2 are the same if $\xi(p)$ increases and directed oppositely if $\xi(p)$ decreases. Setting $\eta(p) \equiv \xi(p)$, we conclude the proof of necessity.

Now assume that there exists a homeomorphism $x = \eta(p)$ possessing the properties formulated in the theorem. We shall construct a function $x = \xi(p)$ as follows. If p is a rest point, then we set $\xi(p) = \eta(p)$. But if $p \in (a, b)$ where (a, b) is an interval adjacent to M_1 and $p = g((a+b)/2, t_0)$, then we set

$$\xi(p) = h\left(\frac{c+d}{2}, t_0\right),$$

where $c = \xi(a)$, $d = \xi(b)$.

In view of the compatibility of the direction of motion the mapping $x = \xi(p)$ constructed is a homeomorphism of R^1 onto R^1 and condition (1.6) is satisfied:

$$\xi(g(p, t)) = \xi\left(g\left(\frac{a+b}{2}, t+t_0\right)\right) = h\left(\frac{c+d}{2}, t+t_0\right)$$

$$= h\left(h\left(\frac{c+d}{2}, t_0\right), t\right) = h(\xi(p), t).$$

Consequently the systems $g(p, t)$ and $h(x, t)$ are isomorphic.

THEOREM 2.6. *Every dynamical system on the real line is isomorphic to some dynamical system defined by means of a differential equation.*

Proof. Let $g(p, t)$ be a dynamical system defined on R^1, M its set of rest points and (a, b) an arbitrary interval adjacent to M. Define a function φ on R^1 by setting $\varphi(x) = 0$ on M and

$$\varphi(x) = (x-a)(x-b)\,\mathrm{sgn}\left[\frac{a+b}{2} - g\left(\frac{a+b}{2}, 1\right)\right]$$

in (a, b). It is clear that the function φ defined in this way is continuous on R^1. Denote the dynamical system defined by the differential equation

$$\frac{dx}{dt} = \varphi(x)$$

by $x = h(x_0, t)$. Clearly, $g(p, t)$ and $h(x_0, t)$ have the same set of rest points M and on every interval (a, b) adjacent to M the motions $h(x_0, t)$ and $g(p, t)$ occur in the same direction since the sign of dx/dt coincides with the sign of the difference

$$g\left(\frac{a+b}{2}, 1\right) - \frac{a+b}{2}.$$

By Theorem 1.6, the dynamical systems $g(p, t)$ and $h(x_0, t)$ are isomorphic. Thus Theorem 2.6 is proved.

At this time it is not known if the analogous theorem is valid in R^2.

Chapter II

LIMITING PROPERTIES OF DYNAMICAL SYSTEMS

§ 7 Dynamical limit points. Properties of limit sets

Definition 1.7. A point q is called an $\omega(\alpha)$-*limit point* of the motion $f(p, t)$ if there exists a sequence $\{t_n\}$ of values of time which tends to $+\infty(-\infty)$ such that the corresponding sequence of images $\{f(p, t_n)\}$ of the point p tends to q.

ω- and α-limit points of the motion $f(p, t)$ are also called *dynamical limit points* of this motion.

THEOREM 1.7. *A necessary and sufficient condition for a point q to be an $\omega(\alpha)$-limit point of the motion $f(p, t)$ is that for any real number $\varepsilon > 0$ and arbitrary time T there exists a time $t > T (t < T)$ such that*

$$f(p, t) \in S(q, \varepsilon). \tag{1.7}$$

Proof. Let q be an ω-limit point of the motion $f(p, t)$. Then there exists a sequence $\{t_n\} \to +\infty$ such that $\{f(p, t_n)\} \to q$. If $\varepsilon > 0$ and T are given, then there exists a natural number N such that $t_N > T$ and $\rho(f(p, t_N), q) < \varepsilon$, i.e., the required condition is satisfied.

Now assume that for any $\varepsilon > 0$ and T there exists a $t > T$ such that inclusion (1.7) is satisfied. Consider the sequences of real numbers $\{\varepsilon_n\} \to 0$ $(\varepsilon_n > 0)$ and $\{T_n\} \to +\infty$. Corresponding to ε_n and T_n there exists a $t_n > T_n$ such that

$$\rho(f(p, t_n), q) < \varepsilon_n. \tag{2.7}$$

Clearly, $\{t_n\} \to +\infty$, and according to inequality (2.7) the sequence $\{f(p, t_n)\} \to q$. Thus q is an ω-limit point of the motion $f(p, t)$.

Definition 2.7. The set of all $\omega(\alpha)$-limit points of the motion $f(p, t)$

is denoted by $\Omega_p(A_p)$ and is called the $\omega(\alpha)$-*limit set* of this motion. The sets Ω_p and A_p are called *the dynamical limit sets* of the motion $f(p, t)$.

We denote the set of all dynamical limit points of the motion $f(p, t)$ by Δ_p. Thus, $\Delta_p = A_p \cup \Omega_p$. It is easy to see that the dynamical limit points of the motion $f(p, t)$ are limit points of the semitrajectories $f(p, R^+)$ and $f(p, R^-)$, i.e.,

$$\Omega_p \subseteq \Sigma_p^+, A_p \subseteq \Sigma_p^-. \tag{3.7}$$

Furthermore, it is not difficult to show that

$$\Sigma_p^+ = f(p, R^+) \cup \Omega_p, \tag{4.7}$$

$$\Sigma_p^- = f(p, R^-) \cup A_p, \tag{5.7}$$

$$\Sigma_p = f(p, R) \cup \Delta_p. \tag{6.7}$$

THEOREM 2.7.   *The dynamical limit sets of the motion $f(p, t)$ are closed.*

Proof. We shall carry out the proof for the set Ω_p. Let q be an arbitrary point of the set $\bar{\Omega}_p$. Then the set Ω_p contains a sequence of points $\{q_n\} \to q$. We shall prove that q is an ω-limit point of the motion $f(p, t)$.

Take an arbitrary ε_0-neighborhood of the point q. Clearly, there exists a point $q_N \in X$ such that $q_N \in S(q, \varepsilon_0)$. There also exists a neighborhood $S(q_N, \varepsilon_1)$ which lies entirely in $S(q, \varepsilon_0)$: $S(q_N, \varepsilon_1) \subseteq S(q, \varepsilon_0)$.

Since the point $q_N \in \Omega_p$, for any given number T there exists a $t > T$ such that $f(p, t) \in S(q_N, \varepsilon_1)$ (see Theorem 1.7). Then $f(p, t) \in S(q, \varepsilon_0)$ and, by Theorem 1.7 again, $q \in \Omega_p$. This completes the proof of the theorem.

THEOREM 3.7.   *The dynamical limit sets of a motion $f(p, t)$ are invariant.*

Proof. We shall prove, for instance, the invariance of the set Ω_p. Consider an arbitrary point $q \in \Omega_p$. There exists a sequence of time values $\{t_n\} \to +\infty$ such that the corresponding sequence of points $\{f(p, t_n)\} \to q$. But then, for arbitrary $t \in R$,

$$\lim_{n \to \infty} (t_n + t) = +\infty,$$

$$\lim_{n \to \infty} f(p, t_n + t) = \lim_{n \to \infty} f(f(p, t_n), t) = f(q, t).$$

Thus, we have found a sequence of time values $\{t_n + t\} \rightarrow +\infty$ for which the corresponding sequence of points $\{f(p, t_n + t)\} \rightarrow f(q, t)$. Consequently, the point $f(q, t) \in \Omega_p$ for any t, i.e., the entire trajectory $f(q, R) \subseteq \Omega_p$ and Ω_p is invariant (see Theorem 2.4).

THEOREM 4.7. *If* $r \in f(p, R)$, *then* $\Omega_r = \Omega_p$.
    Proof. In fact, if $r = f(p, t_0)$, then the assertions $\{f(p, t_n)\} \rightarrow q$ and $\{f(r, t_n - t_0)\} \rightarrow q$ are equivalent and this asserts that $q \in \Omega_p$ if and only if $q \in \Omega_r$.

THEOREM 5.7. *A homomorphism of dynamical systems maps* $\omega(\alpha)$-*limit points into* $\omega(\alpha)$-*limit points*.
    Proof. Let $q \in \Omega_p$, $\xi(p) = x$, $\xi(q) = y$. There exists a sequence of numbers $\{t_n\} \rightarrow +\infty$ such that $\{g(p, t_n)\} \rightarrow q$. Then, according to (1.6),

$$\{h(x, t_n)\} = \{h(\xi(p), t_n)\} = \{\xi(g(p, t_n))\} \rightarrow \xi(q) = y,$$

and consequently $y \in \Omega_x$.

We shall now describe the structure of the dynamical limit sets of various types of motions.

1. If $f(p, t) \equiv p$, i.e., p is a rest point, then the sequence $\{f(p, t)\}$ is the stationary sequence $\{p\}$ and, consequently, it tends to p, whatever the sequence $\{t_n\} \rightarrow +\infty (-\infty)$ is. Therefore, $\Omega_p = A_p = \{p\}$.

2. For a periodic motion $f(p, t)$ with period τ, we have, for any fixed t, that

$$\{t + n\tau\} \rightarrow +\infty, \{t - n\tau\} \rightarrow -\infty$$

and

$$f(p, t \pm n\tau) = f(p, t);$$

therefore the sequence

$$\{f(p, t \pm n\tau)\} \rightarrow f(p, t).$$

Thus, $f(p, t) \in \Omega_p$ and $f(p, t) \in A_p$ for any t, i.e.,

$$f(p, R) \subseteq \Omega_p, \ f(p, R) \subseteq A_p. \tag{7.7}$$

Since the trajectory $f(p, R) = f(p, [0, \tau])$ is compact (see Theorem 1.3), its closure $\Sigma_p = f(p, R)$. Moreover, the inclusions (3.7) can then be written as follows:

$$\Omega_p \subseteq f(p, R), A_p \subseteq f(p, R).$$

These inclusions, together with (7.7), show that for a periodic motion $f(p, t)$, the dynamical limit sets coincide with the trajectory of the motion:

$$\Omega_p = A_p = f(p, R).$$

3. We shall consider the structure of the sets Ω_p and A_p for a non-singular motion $f(p, t)$ in § 11. We now note only that there are nonsingular motions $f(p, t)$ for which the limit

$$\lim_{t \to +\infty} f(p, t) = q(\lim_{t \to -\infty} f(p, t) = q)$$

exists.

Moreover, $\Omega_p = \{q\}(A_p = \{q\})$ and, on the basis of Corollary 1.5, q is a rest point. Thus, in Example 1.1 no motion, except the rest motion, has ω-limit points, and the rest point O is an α-limit point for all these motions. In Example 2.1, O is an ω-limit point of all motions; however, with the exception of the rest point, these motions do not have α-limit points. In Example 3.6, O is an ω-limit point of the motion occurring on the negative half-x-axis and an α-limit point of the motion occurring on the positive half-x-axis.

§ 8   Lagrange stability

Definition 1.8.   A point p and the motion $f(p, t)$ are said to be *positively (negatively) Lagrange stable* if the closure $\Sigma_p^+(\Sigma_p^-)$ of the semitrajectory $f(p, R^+)(f(p, R^-))$ is a compact set. [Note that a necessary and sufficient condition for this is that $f(p, R^+)(f(p, R^-))$ be a sequentially compact set in X.] If, however, the point p and the motion $f(p, t)$ are simultaneously positively and negatively Lagrange stable (Σ_p is compact), then they are said to be *Lagrange stable*.

It follows from the definition that rest points and also periodic motions are Lagrange stable. It is also clear that if the space X is compact, then all motions are Lagrange stable.

In n-dimensional space R^n, the Lagrange stable motions are precisely those motions whose trajectories lie in a bounded subset of the space R^n. It is clear that on the real line all motions, except those occurring on the infinite intervals $(-\infty, a)$, $(b, +\infty)$, $(-\infty, +\infty)$, are Lagrange stable.

LEMMA 1.8.   *The $\omega(\alpha)$-limit set of a motion that is positively (negatively) Lagrange stable is nonempty.*

Proof. Assume the motion $f(p, t)$ is positively Lagrange stable. Take a sequence of positive numbers $\{t_n\} \to \infty$ and consider the corresponding sequence $\{f(p, t_n)\}$. Since

$$f(p, t_n) \in f(p, R) \subseteq \Sigma_p^+,$$

and Σ_p^+ is compact inasmuch as the motion $f(p, t)$ is positively Lagrange stable, the sequence $\{f(p, t_n)\}$ contains a convergent subsequence $\{f(p, t_{n_i})\}$. Suppose $\{f(p, t_{n_i})\} \to q$. Since, moreover, $\{t_{n_i}\} \to +\infty$, q is an ω-limit point of the motion $f(p, t)$. Consequently, the set Ω_p is not empty.

Note that the statement converse to that just proved is not valid: there exist motions which are not positively Lagrange stable whose ω-limit set is not empty.

Example 1.8.   Consider the dynamical system whose motions have the form shown in Fig. 7. [See V. V. Nemytsky and V. V. Stepanov [1], 1949 Russian edition, p. 360, Example 1, or the English edition, p. 341.]

In this example, the motion on spirals is not positively Lagrange stable; however, its ω-limit set is not empty. It consists of the points on the line $x = -1$.

LEMMA 2.8.   *The $\omega(\alpha)$-limit set of motions which are positively (negatively) Lagrange stable is compact.*

This follows from the fact that inclusions (3.7) hold and the compactness of $\Sigma_p^+(\Sigma_p^-)$.

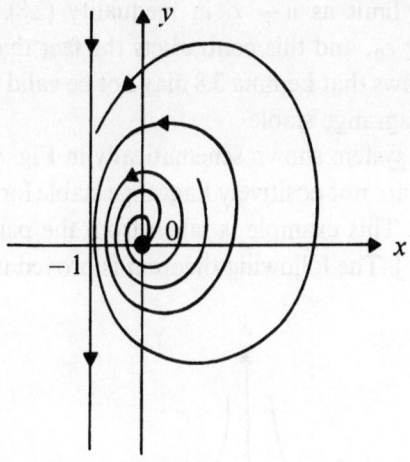

Figure 7

LEMMA 3.8.   *If the motion $f(p, t)$ is positively (negatively) Lagrange stable, then as $t \to +\infty$ ($t \to -\infty$) the point $f(p, t)$ tends to the $\omega(\alpha)$-limit set of this motion.*

Proof. Assume the motion $f(p, t)$ is postively Lagrange stable. Then, by Lemma 1.8, the set Ω_p is not empty. It is required to prove that, as $t \to +\infty$, the distance $\rho(f(p, t), \Omega_p) \to 0$, i.e., that for any $\varepsilon > 0$ there exists a time T such that $\rho(f(p, t), \Omega_p) < \varepsilon$ for all $t > T$.

Suppose this is not true. Then there exists an $\varepsilon_0 > 0$ such that no matter what the number T is there exists a $t > T$ such that

$$\rho(f(p, t), \Omega_p) \geqq \varepsilon_0. \qquad (1.8)$$

Take an arbitrary sequence of positive numbers $\{T_n\} \to +\infty$. Then, according to (1.8), for every natural number n there exists a $t_n > T_n$ such that

$$\rho(f(p, t_n), \Omega_p) \geqq \varepsilon_0. \qquad (2.8)$$

The sequence $\{f(p, t_n)\}$ may not have a limit; however, since Σ_p^+ is compact, this sequence contains a convergent subsequence. In order not to complicate the notation, assume that the sequence $\{f(p, t_n)\}$ converges, and denote its limit by q. Since $\{t_n\} \to +\infty$, $q \in \Omega_p$.

Passing to the limit as $n \to \infty$ in inequality (2.8), we arrive at the inequality $\rho(q, \Omega_p) \geq \varepsilon_0$, and this contradicts the fact that $q \in \Omega_p$.

Example 1.8 shows that Lemma 3.8 may not be valid for motions which are not positively Lagrange stable.

The dynamical system shown schematically in Fig. 8 shows that there exist motions which are not positively Lagrange stable for which Lemma 3.8 is nonetheless valid. This example is taken from the paper by Bronshtein and Shcherbakov [1]. The following theorem is proved in this same paper.

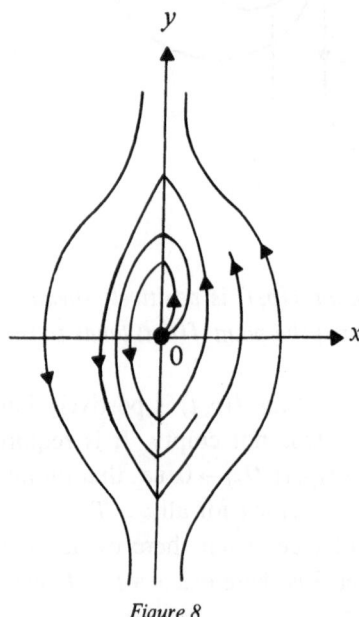

Figure 8

THEOREM 1.8.   *A necessary and sufficient condition for a motion $f(p, t)$ to be positively Lagrange stable is that the following three conditions be satisfied:*

1.8 $\Omega_p \neq \phi,$

2.8 Ω_p *is compact,*

3.8 $\lim\limits_{t \to +\infty} \rho(f(p, t), \Omega_p) = 0.$

Proof. The necessity of these conditions was established in Lemmas 1.8, 2.8 and 3.8. We shall now prove the sufficiency. Thus, assume that conditions 1.8, 2.8 and 3.8 are satisfied. We shall show that then $f(p, R^+)$ is sequentially compact in X. Take an arbitrary sequence $\{q_n\}$ of points $q_n \in f(p, R^+)$. Let $q_n = f(p, t_n)$, where $t_n \geq 0$.

If the sequence $\{t_n\}$ is bounded, then it contains a convergent subsequence and hence the corresponding sequence of points q_n will also converge.

Assume now that $\{t_n\} \to +\infty$. Then, according to condition 3.8,

$$\lim_{n \to \infty} \rho(q_n, \Omega_p) = 0.$$

Therefore, there exists a sequence $\{r_n\}$ of points $r_n \in \Omega_p$ such that

$$\lim_{n \to \infty} \rho(q_n, r_n) = 0. \tag{3.8}$$

By virtue of condition 2.8, the sequence $\{r_n\}$ contains a convergent subsequence. But then, in virtue of (3.8), the corresponding subsequence of points q_n will have the same limit. This completes the proof of the theorem.

It turns out that in locally compact spaces, i.e., in spaces such that for each point there exists a neighborhood with compact closure – for example, n-dimensional Euclidean space – condition 3.8 in Theorem 1.8 is superfluous. To prove this we first establish the following lemma.

LEMMA 4.8. *In a locally compact space X, for every compact set $Q \subseteq X$ there exists an $\varepsilon > 0$ such that $\bar{S}(Q, \varepsilon)$ is compact.*

Proof. In view of the local compactness of X, for each point $q \in Q$ there exists a neighborhood $U(q)$ with compact closure. Since the set Q is compact, this collection of neighborhoods contains a finite subcollection which covers the entire set Q. Let these neighborhoods be $U(q_i)(i = 1, 2, \ldots, m)$. Then

$$U(Q) = \bigcup_{i=1}^{m} U(q_i)$$

is an open set containing Q, i.e., it is a neighborhood of the set Q. Moreover,

$$\overline{U(Q)} = \bigcup_{i=1}^{m} \overline{U(q_i)}$$

is compact. It remains to show that there exists an $\varepsilon > 0$ such that

$$S(Q, \varepsilon) \subseteq U(Q). \tag{4.8}$$

Then $\bar{S}(Q, \varepsilon)$, being a closed subset of the compact set $\overline{U(Q)}$, is compact. Let us assume that, whatever $\varepsilon > 0$ is, inclusion (4.8) is not satisfied, and take a sequence of positive numbers $\{\varepsilon_n\}$ such that $\{\varepsilon_n\} \to 0$. Then there exists a point $r_n \in S(Q, \varepsilon_n)$ such that $r_n \notin U(Q)$. For each point r_n there exists a point $q_n \in Q$ such that

$$\rho(r_n, q_n) < \varepsilon_n. \tag{5.8}$$

Inasmuch as the set Q is compact, the sequence $\{q_n\}$ can be considered to be convergent to some point $q_0 \in Q$. Then, in virtue of (5.8), also $\{r_n\} \to q_0$, and this contradicts the fact that $r_n \notin U(Q)$. This completes the proof of the lemma.

We can now prove the following theorem.

THEOREM 2.8.   *A necessary and sufficient condition for a motion $f(p, t)$ to be positively Lagrange stable in a locally compact space X is that the set Ω_p be nonempty and compact.*

Proof. To prove this theorem it is necessary only to establish that in a locally compact space X conditions 1.8 and 2.8 of Theorem 1.8 imply condition 3.8. Assume that X is locally compact, Ω_p is nonempty and compact, and that condition 3.8 is not satisfied. Then there exists an $\varepsilon_0 > 0$ such that whatever the number T_n is there exists a $t_n > T_n$ such that inequality (2.8) is satisfied. The numbers T_n can be taken so that $\{T_n\} \to +\infty$. Then also $\{t_n\} \to +\infty$. On the other hand, there exists a sequence $\{t'_n\} \to +\infty$ such that

$$\rho(f(p, t'_n), \Omega_p) < \varepsilon_0. \tag{6.8}$$

It follows from (2.8) and (6.8) by virtue of the continuity of the function $\rho(f(p, t), \Omega_p)$ with respect to t that there exists a number τ_n between t_n and t'_n such that

$$\rho(f(p, \tau_n), \Omega_p) = \varepsilon_0. \tag{7.8}$$

On the basis of Lemma 4.8 the number ε_0 can be assumed small enough that the closure of the neighborhood $S(\Omega_p, 2\varepsilon_0)$ is compact. Then the sequence $\{f(p, \tau_n)\}$ can be assumed to be convergent. Let $\{f(p, \tau_n)\} \to r$. According to Lemma 2.2 passage to the limit in equality (7.8) yields $\rho(r, \Omega_p) = \varepsilon_0$.

On the other hand, in virtue of the fact that $\{\tau_n\} \to +\infty$, the point $r \in \Omega_p$. The contradiction thus obtained justifies the assertion of the theorem.

THEOREM 3.8. *If the motion $f(p, t)$ is positively (negatively) Lagrange stable then its $\omega(\alpha)$-limit set is connected.*

[A closed set M is *connected* if it is impossible to represent it as the union of two nonempty closed disjoint sets.]

Proof. Assume that for the positively Lagrange stable motion $f(p, t)$ the set Ω_p is not connected. Then it can be represented in the form of the union of two nonempty closed disjoint sets: $\Omega_p = A \cup B$. Since Ω_p is compact, the distance $\rho(A, B) = d$ between the sets A and B will in this case be positive. [The distance between the sets A and B is understood to be

$$\inf_{a \in A, \, b \in B} \rho(a, b).$$

If $d = 0$ there would exist sequences of points $\{a_n\}$ in A and $\{b_n\}$ in B such that $\rho(a_n, b_n) \to 0$. Inasmuch as $A \cup B$ is compact, the sequences $\{a_n\}$ and $\{b_n\}$ can be assumed to be convergent. Let $\{a_n\} \to a_0, \{b_n\} \to b_0$. Then, since the sets A and B are closed, $a_0 \in A, b_0 \in B$. Furthermore, $\rho(a_0, b_0) = \lim_{n \to \infty} \rho(a_n, b_n) = 0$, and hence $a_0 = b_0$ and $A \cap B \neq \phi$, which contradicts the fact that A and B are disjoint.] We consider two points $r \in A$ and $r' \in B$. Since both of the points r and r' are ω-limit points for the motion $f(p, t)$, there exist two sequences of values of time $\{t_n\} \to +\infty$ and $\{t'_n\} \to +\infty$ such that $\{f(p, t_n)\} \to r$ and $\{f(p, t'_n)\} \to r'$. Then, starting with some n, the inequalities

$$\rho(f(p, t_n), r) < \frac{d}{2}, \rho(f(p, t'_n), r') < \frac{d}{2} \tag{8.8}$$

will hold. Furthermore,

$$\rho(f(p, t_n), A) \leqq \rho(f(p, t_n), r) < \frac{d}{2}. \tag{9.8}$$

On the other hand, taking into consideration the triangle inequality and (8.8), we obtain that

$$d = \rho(A, B) \leqq \rho(A, r') \leqq \rho(A, f(p, t'_n)) + \rho(f(p, t'_n), r')$$

$$< \rho(A, f(p, t'_n)) + \frac{d}{2}.$$

It follows from this that

$$\rho(f(p, t'_n), A) > \frac{d}{2}. \tag{10.8}$$

It follows now from (9.8) and (10.8) by Lemma 2.2 – in virtue of the continuity of the function $\rho(f(p, t), A)$ in t – that there exists a number τ_n between t_n and t'_n such that

$$\rho(f(p, \tau_n), A) = \frac{d}{2}. \tag{11.8}$$

Since the motion $f(p, t)$ is positively Lagrange stable, the sequence $\{f(p, \tau_n)\}$ can be considered to be convergent. Let

$$\lim_{n \to \infty} f(p, \tau_n) = q. \tag{12.8}$$

It follows from (11.8) that $\rho(q, A) = d/2$. Since

$$d = \rho(A, B) \leq \rho(A, q) + \rho(q, B) = \frac{d}{2} + \rho(q, B),$$

we have that $\rho(q, B) \geq d/2$. Thus $q \notin A$ and $q \notin B$ and therefore $q \notin \Omega_p$. At the same time, since $\{\tau_n\} \to +\infty$, it follows from (12.8) that $q \in \Omega_q$. The contradiction thus obtained proves the theorem.

The example depicted in Fig. 7 shows that an ω-limit set may also be connected in the case of a motion which is not positively Lagrange stable.

Fig. 9 gives an example of an ω-limit set (consisting of two parallel lines) which is not connected.

THEOREM 4.8. *All motions in the $\omega(\alpha)$-limit set of a positively (negatively) Lagrange stable motion are Lagrange stable.*

Proof. Assume that the motion $f(p, t)$ is positively Lagrange stable and let the point $q \in \Omega_p$. In virtue of the invariance of Ω_p, the trajectory $f(q, R) \subseteq \Omega_p$ and since Ω_p is closed the set $\Sigma_q \subseteq \Omega_p$. According to Lemma 2.8, the set Ω_p is compact. Then Σ_q is also compact and the motion $f(q, t)$ is Lagrange stable. This completes the proof of the theorem.

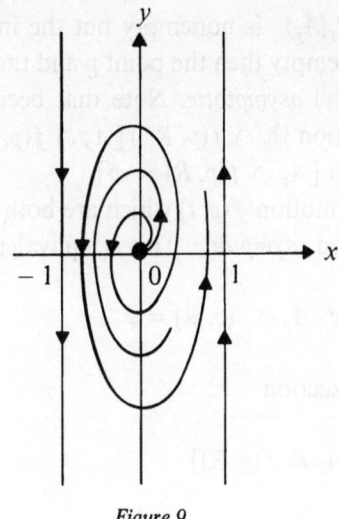

Figure 9

THEOREM 5.8. *A homomorphism of dynamical systems carries positively (negatively) Lagrange stable motions into positively (negatively) Lagrange stable motions.*

Proof. Suppose, for instance, that the motion $g(p, t)$ is positively Lagrange stable. Then the set $g(p, R^+)$ is sequentially compact in X_1. According to Lemma 2.6, $h(p, R^+)$ is the image of the set $g(p, R^+)$ under the continuous mapping $\xi: X_1 \to X_2$. We shall prove that $h(p, R^+)$ is sequentially compact in X_2. Let $\{a_n\}$ be a sequence of points $q_n \in h(p, R^+)$ and let $p_n \in g(p, R^+)$ be such that $q_n \in \xi(p_n)$. The sequence $\{p_n\}$ contains a subsequence $\{p_{n_k}\}$ which converges in X_1 to some point p_0. Then the subsequence $\{a_{n_k}\} = \{\xi(p_{n_k})\}$ converges in X_2 to $\xi(p_0)$.

§ 9   The classification of motions according to the properties of dynamical limit sets

We classify motions $f(p, t)$, starting from properties of dynamical limit points, in the following way:

1. If the set $\Omega_p(A_p)$ is empty then the point p and the motion $f(p, t)$ are called *positively (negatively) departing*. A point p and the motion $f(p, t)$ which are both positively and negatively departing are called *departing* (this is equivalent to the condition $\Delta_p = \phi$).

2. If the set $\Omega_p(A_p)$ is nonempty but the intersection $\Omega_p \cap f(p, R^+)$ $[A_p \cap f(p, R^-)]$ is empty then the point p and the motion $f(p, t)$ are called *positively (negatively) asymptotic*. Note that because of the invariance of $\Omega_p(A_p)$ the intersection $\Omega_p \cap f(p, R^+)[A_p \cap f(p, R^-)]$ is empty if and only if $\Omega_p \cap f(p, R) = \phi \,[A_p \cap f(p, R) = \phi]$.

A point p and motion $f(p, t)$ which are both positively and negatively asymptotic are called *asymptotic*. This is equivalent to

$$\Omega_p \neq \phi, A_p \neq \phi, \Delta_p \cap f(p, R) = \phi.$$

3. If the intersection

$$\Omega_p \cap f(p, R)[A_p \cap f(p, R)]$$

is nonempty then the point p and the motion $f(p, t)$ are called *positively (negatively) Poisson stable*. This case is characterized by the condition that $f(p, R) \subseteq \Omega_p \,[f(p, R) \subseteq A_p]$, and, since the dynamical limit points are closed, by $\Sigma_p \subseteq \Omega_p (\Sigma_p \subseteq A_p)$. Taking (3.7) into consideration, it follows from this that

$$\Omega_p = \Sigma_p (A_p = \Sigma_p). \tag{1.9}$$

A point p and a motion $f(p, t)$ which are both positively and negatively Poisson stable are called *Poisson stable*. A necessary and sufficient condition for a motion $f(p, t)$ to be Poisson stable is that $\Sigma_p = A_p = \Omega_p$. From this it is clear that rest points and periodic motions are always Poisson stable.

It is not difficult to note that for every dynamical system the set of positively (negatively) departing points, the set of positively (negatively) asymptotic points, and the set of positively (negatively) Poisson stable points are invariant and retain these properties under isomorphisms of dynamical systems. This assertion follows directly from Theorems 2.3, 4.7 and 5.7 and Remark 1.6.

Under a homomorphism of dynamical systems a departing motion may go over into an asymptotic, or even into a Poisson stable, motion (e.g., into a rest point). However, the following lemma holds.

LEMMA 1.9.   *Under a homomorphism of dynamical systems, a positively (negatively) Poisson stable motion goes over into a positively (negatively) Poisson stable motion.*

The proof follows from Theorem 5.7.

Example 1.9.   For the dynamical system depicted schematically in Fig. 10, the points O and O_1 are rest points and consequently they are Poisson stable. The motions $f(p, t)$ and $f(r, t)$ are asymptotically stable: $\Omega_p = A_p = O_1, \Omega_r = L(= \text{the circle}), A_r = O$. The motion $f(q, t)$ is positively asymptotically stable $(\Omega_q = L)$ and negatively departing.

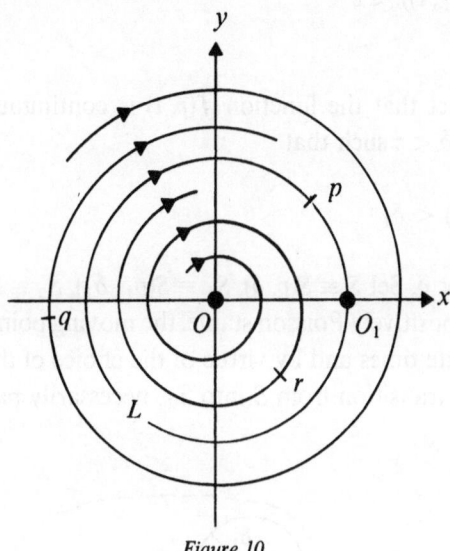

Figure 10

Clearly, the only Poisson stable points on the real line are rest points. In the plane, also the periodic motions are Poisson stable.

THEOREM 1.9.   *For a dynamical system defined in the plane R^2 the singular motions constitute all of the Poisson stable motions.*

Proof. Assume that there exists a nonsingular positively Poisson motion $f(p, t)$ in the plane R^2. Let $0 < t_1 < t_2$. Let $p_i = f(p, t_i)(i = 1, 2)$. Take an arbitrary positive number

$$\varepsilon < \min\left(\frac{d_1}{2}, \frac{d_2}{2}\right),$$

where

41

$$d_1 = \rho(p_1, p_2), d_2 = \rho(p, \overrightarrow{p_1 p_2}).$$

By virtue of the integral continuity, there exists a positive $\delta_1 < \varepsilon$ such that if

$$\rho(r, p_1) < \delta_1 \quad \text{and} \quad 0 < t \leqq t_2 - t_1$$

then the inequality

$$\rho(f(r, t), f(p_1, t)) < \varepsilon$$

is satisfied.

Using the fact that the function $f(p, t)$ is continuous, we can find a positive number $\delta < \varepsilon$ such that

$$\rho(f(q, t_1), p_1) < \delta_1$$

provided $\rho(q, p) < \delta$. Set $S \equiv S(p, \delta)$, $S_1 \equiv S(p_1, \delta_1)$, $S_2 \equiv S(p_2, \varepsilon)$. Since the motion $f(p, t)$ is positively Poisson stable, the moving point $f(p, t)$ will enter S at arbitrarily late times and by virtue of the choice of the numbers δ and δ_1 it will effect a transition from S into S_2, necessarily passing through S_1 in doing so.

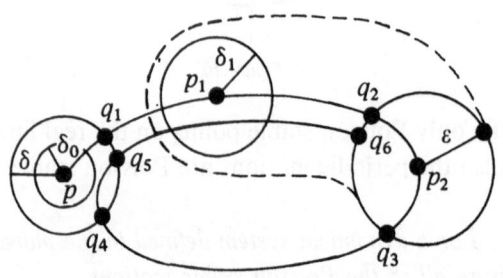

Figure 11

Let $\overset{\frown}{q_1 p_1 q_2}$, $\overset{\frown}{q_3 q_4}$ and $\overset{\frown}{q_5 q_6}$ be three consecutive (in time) arcs of the trajectory $f(p, R)$ which connect S with S_2, S_2 with S and S with S_2 respectively. Each pair of these arcs, together with the circles S and S_2, form a "dumbbell-shaped" curve and furthermore one of these "dumbbells," say M, contains the remaining two. By virtue of the choice of δ_1, the transition of the point $f(p, t)$ as time t increases from S_2 into S through the "handle" of the dumbbell $q_1 q_2 q_5 q_6$ is impossible. Therefore, if M_1 is a handle of the dumbbell M then $M_1 \neq q_1 q_2 q_5 q_6$. Then $M_1 = q_1 q_2 q_3 q_4$ (Fig. 11) or $M_1 = q_3 q_4 q_5 q_6$ (Fig. 12).

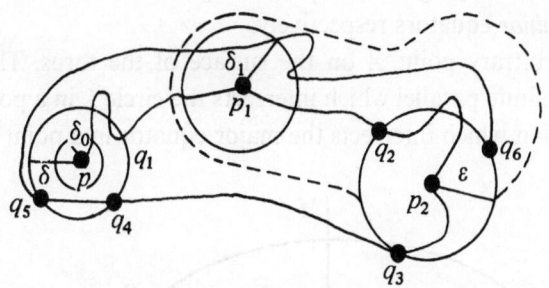

Figure 12

Choose a $\delta_0 > 0$ so that $S(p, \delta_0) \cap \widehat{q_2 q_5} = \phi$. It is obvious that in both cases the semitrajectory $f(q_6, R^+)$ cannot enter $S(p, \delta_0)$ since for this to happen it must either intersect the arc $\widehat{pq_6}$ of the trajectory $f(p, R)$ or pass from S_2 into S through the handle of the dumbbell $q_1 q_2 q_5 q_6$. Neither of these situations is possible. This completes the proof of the theorem.

The idea behind the proof given here is due to A. F. Filippov [1] (page 122). An analogous proof was published in the paper by P. Seibert and P. Tulley [1]. A more complicated proof was proposed by H. Bohr and W. Fenchel [1]. For dynamical systems defined by a system of differential equations in the plane, this theorem was first proved by I. Bendixson [1].

We shall give below examples of Poisson stable trajectories which are different from a rest point and from closed curves. Such trajectories may occur for instance on the torus.

§ 10 Examples of Poisson stable motions on the torus

Example 1.10. Consider the torus, i.e., the surface obtained by rotating a circle L about an axis MM' lying in the plane of this circle and not intersecting it (see Fig. 13). The line MM' is called the *axis* of the torus. As the circle L describes the torus upon rotation about the axis MM', its center O_1 moves along a circle called the *axial* circle of the torus. Its center O is called the *center* of the torus and the plane of the axial circle of the torus is called the *equatorial plane* of the torus. The circles obtained by intersecting the torus with planes which pass through the axis MM' are called the *meridians* of the torus and the circles obtained by intersecting the torus with planes perpendicular to the axis MM' are called *parallels* (of latitude) of the torus. Parallels lying in the equatorial plane of the torus are called the *major* and *minor* equators respectively.

Take an arbitrary point A on the surface of the torus. Through A there passes a definite parallel which intersects the circle L in a point B and a definite meridian which intersects the major equator in a point C.

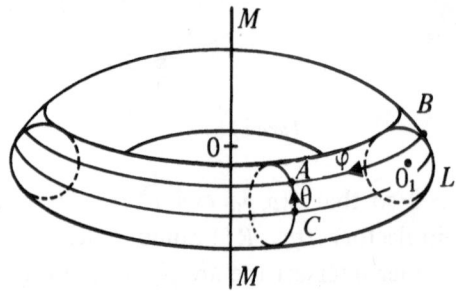

Figure 13

Denote the radian measures of the arcs BA and CA reckoned in the directions indicated in Fig. 13 by φ and θ respectively.

For arbitrary integer values of m and n, the numbers $\varphi + 2\pi m$ and $\theta + 2\pi n$ are called the coordinates of the point A (they are the longitudes and latitudes respectively).

If $\{\varphi_1\}$ and $\{\varphi_2\}$ are the sets of longitudes of the points A_1 and A_2 respectively on the torus and $\{\theta_1\}$ and $\{\theta_2\}$ are the sets of latitudes of these points then we agree to let the *distance* between A_1 and A_2 be $\sqrt{\varphi^2 + \theta^2}$ where

$$\varphi = \inf\{|\varphi_2 - \varphi_1|\}, \ \theta = \inf\{|\theta_2 - \theta_1|\}.$$

Clearly, the maximal distance between points on the torus equals

$$\sqrt{\pi^2 + \pi^2} = \sqrt{2}\pi.$$

Now define a dynamical system on the torus by means of a system of differential equations

$$\frac{d\varphi}{dt} = 1, \quad \frac{d\theta}{dt} = \alpha, \tag{1.10}$$

where $\alpha > 0$ is a real number. Integrating this system with the initial data $\varphi = \varphi_0, \theta = \theta_0$ for $t = 0$, we obtain:

$$\varphi = \varphi_0 + t, \theta = \theta_0 + \alpha t. \tag{2.10}$$

The equality $\varphi = \varphi_0 + t$ shows that the moving point (φ, θ) rotates in a uniform manner about the axis of the torus, performing thereby an infinite number of rotations. It is clear from the equality $\theta = \theta_0 + \alpha t$ that the moving point (φ, θ) also rotates about the axial circle of the torus in a uniform manner. The equation of the trajectory of the motion can be written as

$$\theta - \theta_0 = \alpha(\varphi - \varphi_0).$$

One must consider those points of the torus to belong to the trajectory for which at least one pair of coordinates satisfies this equation. It is not difficult to see that this trajectory can be obtained from the trajectory $\theta = \alpha\varphi$ if for each point of the latter one increases the longitude by $\varphi_0 - \theta_0/\alpha$ (i.e., if one rotates the torus about its axis through the angle $\varphi_0 - \theta_0/\alpha$). Because of this we limit our investigation to the trajectories $\theta = \alpha\varphi$, setting $\varphi_0 = \theta_0 = 0$ in (2.10).

Two cases are possible:

1. The number α is rational: $\alpha = p/q$, where p and q are relatively prime natural numbers. Then $\theta = (p/q)\varphi$ and when φ increases by the amount $2\pi q$ then the latitude θ increases by the amount $2\pi p$. Thus, the trajectory closes after q rotations about the axis of the torus (or after p rotations about the axial circle). In this case, according to (2.10), all motions are periodic with period $2\pi q$.

2. The number α is irrational. In this case the motion

$$\varphi = t, \theta = \alpha t \tag{3.10}$$

cannot be periodic. In fact, this motion passes through the point $(0 + 2\pi n, 0 + 2\pi m)$ at $t = 0$. If we assume that it passes through this same point for some value $t_0 > 0$ then we obtain that for some natural numbers m and n

$$\varphi = t_0 = 2\pi n, \theta = \alpha t_0 = 2\pi m,$$

and hence $2\pi n\alpha = 2\pi m$ and the number $\alpha = m/n$ must be rational.

Consequently in this case all motions are non-periodic. We shall now prove that they are Poisson stable.

To this end, we limit ourselves to considering the motions (3.10) for $t > 0$ and we study the points of intersection of the positive semitrajectory with an arbitrary meridian $\varphi = \varphi_0 > 0$, i.e., the points

$$\{(\varphi_0 + 2\pi n,\ \alpha\varphi_0 + 2\pi n\alpha)\}, \tag{4.10}$$

where n is any natural number. We shall show that these points form an everywhere dense set on the meridian $\varphi = \varphi_0$. To do this it suffices to prove that the numbers $2\pi n\alpha - 2\pi[n\alpha] = 2\pi(n\alpha)$ form an everywhere dense set on the segment $[0, 2\pi]$ or that the numbers $n\alpha - [n\alpha] = (n\alpha)$ are distributed everywhere densely on the segment $[0, 1]$. ($[\beta]$ denotes the greatest integer less than or equal to β and $(\beta) = \beta - [\beta]$.) Now partition the segment $[0, 1]$ into p equal parts (see Fig. 14) and consider the numbers $(\alpha), (2\alpha), \ldots, ((p+1)\alpha)$. Then, since there are more numbers than segments, there exists at least one segment $((h-1)/p, h/p)$ which contains at least two of the numbers under consideration. Suppose these are the numbers $(k_1\alpha)$ and $(k_2\alpha)(k_2 > k_1)$:

$$\frac{h-1}{p} < k_1\alpha - [k_1\alpha] < \frac{h}{p}$$

and

$$\frac{h-1}{p} < k_2\alpha - [k_2\alpha] < \frac{h}{p}.$$

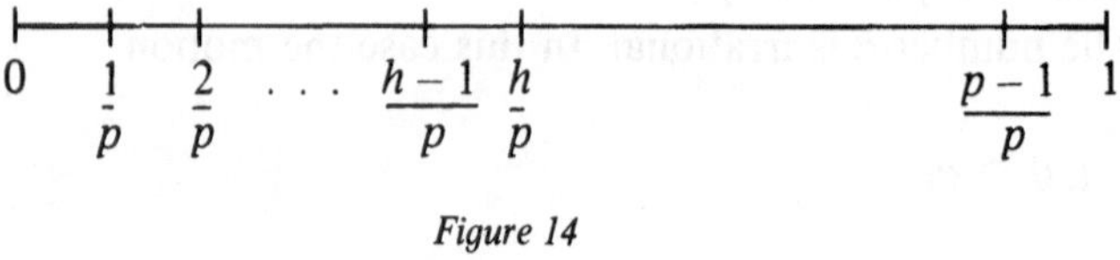

Figure 14

It follows that

$$-\frac{1}{p} < (k_2 - k_1)\alpha - [k_2\alpha] + [k_1\alpha] < \frac{1}{p}.$$

Setting

$$k_2 - k_1 = N, \; [k_2\alpha] - [k_1\alpha] = M,$$

we obtain that

$$-\frac{1}{p} < N\alpha - M < \frac{1}{p}, \quad \text{i.e.,} \quad |N\alpha - M| < \frac{1}{p}.$$

Finally, if we set $|N\alpha - M| = \gamma$, we will have $\gamma < 1/p$. Here two cases are possible:

1. $N\alpha - M > 0$. Then $\gamma = N\alpha - M$ and consequently

$$N\alpha = M + \gamma, \qquad (N\alpha) = \gamma,$$

$$2N\alpha = 2M + 2\gamma, \qquad (2M\alpha) = 2\gamma,$$

$$\cdot \quad \cdot \quad \cdot \quad \cdot \quad \cdot \quad \cdot \quad \cdot \quad \cdot$$

$$lN\alpha = lM + l\gamma, \qquad (lM\alpha) = l\gamma,$$

if furthermore $l\gamma < 1$.

2. $N\alpha - M < 0$. Then $\gamma = M - N\alpha$ and consequently

$$N\alpha = M - \gamma, \qquad (N\alpha) = 1 - \gamma,$$

$$2N\alpha = 2M - 2\gamma, \quad (2N\alpha) = 1 - 2\gamma,$$

$$\cdot \quad \cdot \quad \cdot \quad \cdot \quad \cdot \quad \cdot \quad \cdot \quad \cdot$$

$$lN\alpha = lM - l\gamma, \qquad (lM\alpha) = 1 - l\gamma,$$

if furthermore $l\gamma < 1$.

In the first case the numbers $(N\alpha), (2N\alpha), \ldots, (lN\alpha)$ are distributed on the segment $[0, 1]$ in increasing order and in the second case in decreasing order – at the same distance equal to $\gamma < 1/p$ from one another. Taking into consideration that p can be taken arbitrarily large we conclude that the set of numbers $\{(n\alpha)\}$ is everywhere dense on the segment $[0, 1]$, which is what was required to be proved.

Thus the set of points (4.10) is everywhere dense on the meridian $\varphi = \varphi_0$ and therefore any point (φ_0, θ_0) on this meridian will be a limit point for the sequence (4.10). Clearly, according to (3.10), the sequence of values of time $\{t_n = \varphi_0 + 2\pi n\}$ which correspond to the points (4.10) tend to $+\infty$ and therefore the point (φ_0, θ_0) is an ω-limit point for the motion (3.10). Inasmuch as φ_0 and θ_0 are arbitrary, we conclude that the ω-limit set of the motion (3.10) is the entire torus and in particular there exist ω-limit points belonging to the trajectory, i.e., this motion is positively Poisson stable. It is proved analogously that the motion is negatively Poisson stable and consequently that it is Poisson stable.

Example 2.10.   Consider the dynamical system defined on the torus by means of the differential equations

$$\frac{d\varphi}{dt} = \Phi(\varphi, \theta), \frac{d\theta}{dt} = \alpha\Phi(\varphi, \theta), \tag{5.10}$$

where $\alpha > 0$ is an irrational number and $\Phi(\varphi, \theta)$ is a continuous function on the torus which is positive everywhere except at the point $(0, 0)$ and $\Phi(0, 0) = 0$. Such a function is for example

$$\Phi(\varphi, \theta) = \rho((\varphi, \theta), (0, 0)).$$

Clearly, the trajectories of the motion in this case are like those in the system (1.10) with the exception of the curve $\theta = \alpha\varphi$, passing through the origin $(0, 0)$, which in this case is a rest point. This rest point divides the Poisson stable trajectory $\theta = \alpha\varphi$ of Example 1.10 into three trajectories: a rest point, a positively asymptotic negatively Poisson stable trajectory, and a negatively asymptotic positively Poisson stable trajectory. Moreover, it is clear that all motions slow down upon passing near the point $(0, 0)$ since the speed of motion equals $\Phi(\varphi, \theta)\sqrt{1 + \alpha^2}$.

§ 11   Properties of Poisson stable points and motions

It was pointed out in §9 that a necessary and sufficient condition for a point p and a motion $f(p, t)$ to be positively (negatively) Poisson stable is that equality (1.9) be satisfied, and it was also pointed out that the set of positively (negatively) Poisson stable points is invariant; in view of this one can speak of positively (negatively) Poisson stable trajectories and of Poisson stable trajectories.

THEOREM 1.11. *A point p and a motion $f(p, t)$ are positively (negatively) Poisson stable if and only if the positive semitrajectory $f(p, R^+)$ [the negative semitrajectory $f(p, R^-)$] is not homeomorphic to a half-line.*

Proof. Suppose the nonsingular motion $f(p, t)$ is positively Poisson stable. Then the mapping

$$q = f(p, t) \equiv \varphi(t) \qquad (0 \leq t < +\infty), \tag{1.11}$$

although 1–1 and continuous, will not be a homeomorphism since the inverse mapping

$$t = \varphi^{-1}(q) \qquad (q \in f(p, R^+)) \tag{2.11}$$

is not continuous because in this case $p \in \Omega_p$ and there exists a sequence

$$\{q_n = \varphi(t_n) = f(p, t_n)\} \rightarrow p = \varphi(0)$$

such that $t_n \geq 0$ and

$$\{t_n = \varphi^{-1}(q_n)\} \rightarrow +\infty$$

and not to $\varphi^{-1}(p)$.

Now assume that the positive semitrajectory $Q = f(p, R^+)$ is homeomorphic to the half-line $t_0^* \leq t^* < +\infty$, i.e., there exists a 1–1 continuous mapping

$$q = \psi(t^*) \tag{3.11}$$

of the half-line $t_0^* \leq t^* < +\infty$ onto the curve Q for which the inverse mapping $t^* = \psi^{-1}(q)$ is continuous. Then the mapping

$$t^* = \psi^{-1}(\varphi(t)) \tag{4.11}$$

of the half-line $0 \leq t < +\infty$ onto the half-line $t_0^* \leq t^* < +\infty$ is 1–1 and continuous – therefore, it is a homeomorphism (see §2). Then the mapping (1.11), being the composition of the homeomorphisms (4.11) and (3.11), must be a homeomorphism, which however is not the case. It has thus been proved in the case when the motion $f(p, t)$ is positively Poisson stable that

49

the positive semitrajectory $f(p, R^+)$ is not homeomorphic to a half-line.

Now assume that the motion $f(p, t)$ is not positively Poisson stable. Then the mapping (2.11) must be continuous. In fact, if the sequence $\{q_n\}$ of points on the positive semitrajectory $f(p, R^+)$ tends to some point $q_0 \in f(p, R^+)$ then the corresponding sequence $\{t_n = \varphi^{-1}(q_n)\}$ $(t_n \geq 0)$ is bounded (for otherwise $q_0 \in \Omega_p$ and the motion is positively Poisson stable). Let t_0 be a limit point of the sequence $\{t_n\}$. Then since, according to (1.11),

$$q_n = f(p, t_n) = \varphi(t_n),$$

we have that

$$q_0 = f(p, t_0) = \varphi(t_0).$$

It follows from this that $t_0 = \varphi^{-1}(q_0)$. This means that the limit

$$\lim_{n \to \infty} t_n = t_0 = \varphi^{-1}(q_0)$$

exists, which is what was required to be proved.

Theorem 1.11 immediately implies the following two corollaries.

COROLLARY 1.11. *The point p and the motion $f(p, t)$ are at least either positively or negatively Poisson stable if and only if the trajectory $f(p, R)$ is not homeomorphic to the real line.*

COROLLARY 2.11. *The point p and the motion $f(p, t)$ are neither positively nor negatively Poisson stable if and only if the trajectory $f(p, R)$ is homeomorphic to the real line.*

[Corollary 2.11 was established by R. È. Vinograd [1] (page 99); see also the papers by N. E. Foland [1] (Theorems 1 and 2) and J. W. England [1].]

It is not difficult to see that for motions which are not positively (negatively) Poisson stable we have

$$\Omega_p = \Sigma_p^+ \setminus f(p, R^+)[A_p = \Sigma_p^- \setminus f(p, R^-)],$$

and that the equality

$$A_p = \Sigma_p \setminus f(p, R)$$

holds for motions which are neither positively nor negatively Poisson stable.

As regards positively (negatively) Poisson stable motions, we have, as was remarked above (see (1.9)), that $\Omega_p = \Sigma_p$ $(A_p = \Sigma_p)$, and cases can occur when $\Omega_p(A_p)$ does not contain points other than points of the trajectory through p. This will be the situation, for example, if in Example 2.10 one considers as the space X only the negatively asymptotic positively Poisson stable trajectory. But such a space is not complete.

THEOREM 2.11.   *In a complete space the set of points which do not belong to the trajectory of a nonsingular positively (negatively) Poisson stable motion* $f(p, t)$ *is everywhere dense in the set* $\Omega_p(A_p)$, *i.e.,*

$$\Omega_p = \overline{\Sigma_p \setminus f(p, R)}[A_p = \overline{\Sigma_p \setminus f(p, R)}].$$

Proof. Assume that the space X is complete and that the nonsingular motion $f(p, t)$ is positively Poisson stable. We shall show that for any point $p_0 \in \Omega_p$ and arbitrary $\varepsilon > 0$ there exists a point

$$q \in \Omega_p \cap S(p_0, \varepsilon)$$

such that $q \notin f(p, R)$.

Inasmuch as $p_0 \in \Omega_p$, there exists a $t_1 > 1$ such that

$$f(p, t_1) \in S(p_0, \varepsilon).$$

Set $f(p, t_0) = p_1$. Since the motion $f(p, t)$ is nonsingular and $t_1 > 1$ we have that

$$p_1 \notin f(p, [-1, 1]).$$

Take an ε_1-neighborhood of the point p_1 such that
   1. $S(p_1, \varepsilon_1) \subseteq S(p_0, \varepsilon)$;

   2. $\overline{S}(p_1, \varepsilon_1) \cap f(p, [-1, 1]) = \phi$;

   3. $\varepsilon_1 < \dfrac{\varepsilon}{2}$.

Since

$$p_1 \in f(p, R) \subseteq \Omega_p,$$

there exists a $t_2 > 2$ such that

$$f(p, t_2) \in S(p_1, \varepsilon_1).$$

Set $f(p, t_2) = p_2$. It is clear that

$$p_2 \notin f(p, [-2, 2]).$$

Now take an ε_2-neighborhood of the point p_2 such that

1. $S(p_2, \varepsilon_2) \subseteq S(p_1, \varepsilon_1)$;
2. $\bar{S}(p_2, \varepsilon_2) \cap f(p, [-2, 2]) = \phi$;

3. $\varepsilon_2 < \dfrac{\varepsilon_1}{2}$.

Continuing this process, we obtain a nested sequence of nonempty balls

$$\bar{S}(p_0, \varepsilon) \supseteq \bar{S}(p_1, \varepsilon_1) \supseteq \ldots \supseteq \bar{S}(p_n, \varepsilon_n) \supseteq \ldots$$

which, since the space X is complete and $\{\varepsilon_n\} \to 0$, has nonempty intersection

$$q = \bigcap_{n=1}^{+\infty} \bar{S}(p_n, \varepsilon_n).$$

We shall show that $q \in \Omega_p$ but does not belong to the trajectory $f(p, R)$. It follows from the fact that

$$q \in \bar{S}(p_n, \varepsilon_n) \tag{5.11}$$

that $\rho(p_n, q) \leq \varepsilon_n$, and therefore

$$\lim_{n \to \infty} p_n = q.$$

But $p_n = f(p, t_n)$ whereas

$$\lim_{n \to \infty} t_n = +\infty.$$

This means that $q \in \Omega_p$. Inasmuch as

$$\bar{S}(p_n, \varepsilon_n) \cap f(p, [-n, n]) = \phi,$$

it follows from (5.11) that

$$q \notin f(p, [-n, n]),$$

i.e., $q \notin f(p, R)$, which is what was required to be proved.

Chapter III

NONWANDERING POINTS. CENTRAL MOTIONS

§ 12 Wandering and nonwandering points

Definition 1.12. The point p is called a *nonwandering point in X* or simply a *nonwandering point* if for any $T \in R$ and any neighborhood $U(p) \subseteq X$ of the point p there exists a moment of time $t > T$ such that

$$U(p) \cap f(U(p), t) \neq \phi. \tag{1.12}$$

It is not difficult to prove the next theorem.

THEOREM 1.12. *A point p is a nonwandering point if and only if for every neighborhood $U(p)$ and every $T \in R$ there exists a point $q \in U(p)$ (the neighborhood $U(q) \subseteq U(p)$) and a moment of time $t > T$ such that $f(q, t) \in U(p)$ ($f(U(q), t) \subseteq U(p)$).*

We denote the set of all nonwandering points by M. The points of the complementary set $W = X \setminus M$ are called *wandering* points. The following definition is equivalent to the definition just given.

Definition 2.12. The point p is called a *wandering point in X* or simply a *wandering point* if there exists a number T and a neighborhood $U(p)$ of the point p such that for all $t > T$,

$$U(p) \cap f(U(p), t) = \phi. \tag{2.12}$$

In Example 1.9 (see Fig. 10) both of the rest points O and O_1 and all of the points on the circle L are nonwandering points. The remaining points, i.e., the points on the two spirals, are wandering points. In Example 1.8, the rest point O and the points on the line $x = -1$ are nonwandering points.

Remarks 1.12. Applying a transformation with the parameter $-t$ to the left members of relations (1.12) and (2.12) and taking into consideration that, according to (2.2), the image of an intersection equals the intersection of the images, we find in the first case that $U(p) \cap f(U(p)), -t) \neq \phi$ and in the second case that $U(p) \cap f(U(p), -t) = \phi$. [It is not difficult to show that if $A \cap B = \phi$, then also $f(A, t_0) \cap f(B, t_0) = \phi$. In fact, if $f(A, t_0) \cap f(B, t_0) \neq \phi$, then, applying a transformation with parameter $-t_0$, we obtain that $A \cap B \neq \phi$.] This shows that relation (1.12), being satisfied for some $t - t_0 > T$, is also satisfied for $t = -t_0 < -T$, and that relation (2.12), being satisfied for all $t > T$, is also satisfied for all $t < -T$, i.e., the properties of a point to be wandering or nonwandering are two-sided properties.

THEOREM 2.12. *The set of wandering points is open.*

In fact, it is clear from relation (2.12) that, together with the point $p \in W$, all points in a neighborhood $U(p)$ are also wandering points, and this means that the set W is open.

COROLLARY 1.12. *The set $M = X \setminus W$ of nonwandering points is closed.*

THEOREM 3.12. *The set of wandering points is invariant.*

Proof. Let $p \in W$. Then there exists a number T and a neighborhood $U(p)$ of the point p such that for all $t > T$ the relation (2.12) holds. Applying to this relation a mapping with parameter t_0, we obtain that for all $t > T$,

$$f(U(p), t_0) \cap f(U(p), t + t_0) = \phi,$$

i.e.,

$$f(U(p), t_0) \cap f(f(U(p), t_0), t) = \phi.$$

[See the note in square brackets in Remarks 1.12, above.] Setting

$$f(U(p), t_0) \equiv U_1(f(p, t_0)),$$

we find that for all $t > T$,

$$U_1(f(p, t_0)) \cap f(U_1(f(p, t_0)), t) = \phi. \tag{3.12}$$

The set $U_1(f(p, t_0))$, being according to Theorem 2.2 the topological

55

image of an open set $U(p)$, is an open set which contains the point $f(p, t_0)$ and consequently it is a neighborhood of the point $f(p, t_0)$. In this connection, we conclude from (3.12) that the point $f(p, t_0)$ is a wandering point. Thus $f(p, t_0) \in W$ for any $t_0 \in R$ and consequently $f(p, R) \subseteq W$ and hence the set W is invariant. From this, taking Theorem 4.4 into consideration, we obtain the following corollary.

COROLLARY 2.12.   *The set of nonwandering points is invariant.*

Remark 2.12.   Since for any neighborhood of the point p one can find a spherical neighborhood with center at the point p contained in it, the neighborhood $U(p)$ in Definitions 1.12 and 2.12 can be assumed to be a spherical neighborhood.

§ 13   Properties of the set of nonwandering points

Let us now consider the initial metric space to be an invariant set $A \subseteq X$. If, in this connection, the point $p \in A$ is nonwandering, then we say it is nonwandering in A.

THEOREM 1.13.   *If a point p which belongs to an invariant set $A \subseteq X$ is nonwandering in A, then it is also nonwandering in X.*

Proof.   On the basis of Remark 1.12, the fact that the point is nonwandering in A means that for any T and any ε-neighborhood $U_A(p)$ of the point p there exists a $t > T$ such that

$$U_A(p) \cap f(U_A(p), t) \neq \phi. \tag{1.13}$$

Now consider the ε-neighborhood $U(p)$ of the point p in X. Then $U(p) \cap A = U_A(p)$ is an ε-neighborhood of the point p in A. By virtue of (1.13), relation (1.12) holds *a fortiori* for the $t > T$ whose existence was stipulated, and consequently the point p is nonwandering in X. This completes the proof of the theorem.

One must not think however that every point which is wandering in an invariant set $A \subset X$ is wandering in X. Thus, in Example 1.9, the point p is nonwandering in X and wandering in $f(p, R)$.

THEOREM 2.13.   *Every dynamical limit point q is nonwandering in Σ_p if $q \in \Delta_p$, and in $f(p, R)$ if $q \in \Delta_p \cap f(p, R)$.*

Proof. Assume for definiteness that $q \in \Omega_p$. Then in any neighborhood $S(q, \varepsilon)$ of the point q there exists a point $p_0 \in f(p, R)$. Since, on the basis of Theorem 4.7, $\Omega_{p_0} = \Omega_p$, we have that $q \in \Omega_{p_0}$, and for every $T \in R$ there exists a $t > T$ such that $f(p_0, t) \in S(q, \varepsilon)$. On the basis of Theorem 1.12, the point $q \in M$. Thus $q \in \Sigma_p$ and $p_0 \in f(p, R)$. This completes the proof of the theorem.

Taking into consideration that a point p is at least positively or negatively Poisson stable if and only if $p \in \Delta_p$, we obtain the following corollary to Theorem 2.13.

COROLLARY 1.13.   *Every point p which is at least positively or negatively Poisson stable is nonwandering in $f(p, R)$ and hence also in X.*

It is easy to see that the converses of Theorem 2.13 and Corollary 1.13 do not hold. In fact, if we replace the spirals $f(q, R)$ and $f(r, R)$ in Example 1.9 by families of circles with center at the point O, then the point p will be nonwandering in X – however, p is neither positively nor negatively Poisson stable and it is not even a dynamical limit point.

Remark 1.13.   Given a dynamical system, there need not exist a nonempty set of nonwandering points. Thus, in Example 5.6, all points are wandering points. However, taking Lemma 1.8 and Theorem 2.13 into consideration, we obtain the following corollary.

COROLLARY 2.13.   *If there exists at least one motion which is at least either positively or negatively Lagrange stable, then the set of nonwandering points is nonempty.*

Taking into consideration the fact that in a compact space all motions are Lagrange stable, this implies the next corollary.

COROLLARY 3.13.   *In a compact metric space the set of nonwandering points is nonempty.*

The next corollary also follows from Lemma 3.8 and Theorem 2.13.

COROLLARY 4.13.   *In a compact metric space every motion tends to the set of nonwandering points.*

§ 14   The set of central motions

Throughout this section we shall assume the space X to be *compact*. Then, according to Corollary 3.13, the set M of nonwandering points is non-empty. Moreover, M is invariant and closed (see Corollaries 1.'2 and 2.12). M is compact inasmuch as it is a closed subset of a compact space. If we consider M to be the initial space, then, since it is compact, the set M_1 of nonwandering points in M ($M_1 \subseteq M$) is nonempty. Thus, in Example 1.9 the set M_1 consists solely of two rest points.

Considering M_1 as the initial space, we find a nonempty closed (and hence compact) invariant set M_2 of points which are nonwandering in M_1 ($M_2 \subseteq M_1$). Continuing this process, we obtain an infinite chain of nested closed sets

$$M \supseteq M_1 \supseteq M_2 \supseteq \ldots \supseteq M_k \supseteq \ldots.$$

Their intersection $M_\omega \equiv \bigcap_{k=1}^{\infty} M_k$ is nonempty inasmuch as the space X is compact. [See P. S. Aleksandrov [2], p. 381, property b), according to which the compactness of a topological space is equivalent to the fact that every decreasing sequence of nonempty closed sets has nonempty intersection.] Moreover, M_ω is invariant and closed (and hence compact), as is every intersection of invariant closed sets, and M_ω can be considered as the initial space in which the set $M_{\omega+1}$ of nonwandering points is nonempty ($M_{\omega+1} \subseteq M_\omega$). Continuing this process, we obtain the sequence

$$M \supseteq M_1 \supseteq M_2 \supseteq \ldots \supseteq M_k \supseteq \ldots \supseteq M_\omega \supseteq M_{\omega+1} \supseteq \ldots \supseteq M_{\omega+k} \supseteq \ldots,$$

and then we again take the intersection $M_\alpha \equiv \bigcap_{k=1}^{\infty} M_{\omega+k}$ and denote by $M_{\alpha+1}$ the nonempty set of points which are nonwandering in M_α, and so forth. Thus, we obtain the following transfinite sequence of closed sets

$$M \supseteq M_1 \supseteq \ldots \supseteq M_k \supseteq \ldots \supseteq M_\omega \supseteq \ldots \supseteq M_\alpha \supseteq \ldots \supseteq M_\beta \supseteq \ldots. \quad (1.14)$$

Since a compact (metric) space always possesses a countable basis, by applying the Baire-Hausdorff theorem we arrive at the conclusion that after no more than a countable number of steps the sets in the chain (1.14) begin to coincide, i.e., there exists a γ such that $M_\gamma = M_{\gamma+1} = \ldots.$ [A *basis* for the space X is a system $S = \{U_\sigma\}$ of open sets of the space X

such that any open set $U \subseteq X$ can be written as the union of certain sets which are elements of the system S. A system of neighborhoods $S = \{U_\alpha\}$ is a basis for the space X if and only if for any point $p \in X$ and any neighborhood $U(p)$ of p there exists a neighborhood $U_\sigma \in S$ such that $p \in U_\sigma \subseteq U(p)$. A necessary and sufficient condition for a metric space X to have a countable basis is that it be *separable*, i.e., that X contain an everywhere dense countable subset (see P. S. Aleksandrov [2], pp. 269–270).] [Concerning the Baire-Hausdorff theorem, see P. S. Aleksandrov [2], p. 274, Theorem 53 (this theorem is sometimes called the Cantor-Baire Theorem): If a space with countable basis has a well-ordered sequence (1.14) of nested closed sets, then it contains at most a countable number of distinct elements – recall that an *ordered* set is called *well ordered* if every nonempty subset in it contains a first element.]

Definition 1.14.   The set M_γ is called the *set of central motions* or simply the *center* of the dynamical system.

We denote M_γ by Z. Obviously, Z is a closed invariant set and, furthermore, since $M_{\gamma+1} = M_\gamma$, all the points in Z are nonwandering in Z. It is also clear that the following situation occurs.

Remark 1.14. Z is the maximal closed invariant set consisting entirely of points which are nonwandering in it. In particular, Z contains all points p which are nonwandering in $f(p, R)$. From this, on the basis of Corollary 1.13, we conclude that all points which are at least positively or negatively Poisson stable necessarily appear in the set of central motions.

THEOREM 1.14.   *The set of central motions contains an everywhere dense set of Poisson stable points.*

Proof. Consider the set Z to be the initial space. Take an arbitrary point $p \in Z$ and consider the ε-neighborhood $S \equiv S(p, \varepsilon)$. It is required to prove that S contains at least one Poisson stable point. Since the point p is nonwandering, we have, on the basis of Theorem 1.12 and Remark 2.12, that there exist a $t_1 > 1$ and a ball S_1 such that

$$S_1 \subseteq S, \quad f(S_1, t_1) \subseteq S.$$

If necessary, by decreasing the radius ε_1 of the ball S_1 we can assume that $\varepsilon_1 < \varepsilon/2$ and that

$$\bar{S}_1 \subseteq S, \quad f(\bar{S}_1, t_1) \subseteq S.$$

Taking into consideration that all points in Z are nonwandering in it, we find, again on the basis of Theorem 1.12 and Remark 1.12, for the ball S_1 a $t_2 < -2$ and a ball S_2 of radius $\varepsilon_2 < \varepsilon_1/2$ such that

$$\bar{S}_2 \subseteq S_1, \, f(\bar{S}_2, t_2) \subseteq S_1.$$

Analogously, we find a $t_3 > 3$ and a ball S_3 of radius $\varepsilon_3 < \varepsilon_2/2$ such that

$$\bar{S}_3 \subseteq S_2, \, f(\bar{S}_3, t_3) \subseteq S_2,$$

and so forth. We thus obtain a decreasing nested sequence of closed balls which has the sole common point $q = \bigcap_{n=1}^{\infty} \bar{S}_n$ (since a compact space is complete).

Consider an arbitrary neighborhood $S(q, \delta)$ of the point q. There exists an $\bar{S}_N \subseteq S(q, \delta)$. Then

$$f(\bar{S}_n, t_n) \subseteq S_N \subseteq S(q, \delta)$$

for all $n > N$. In particular, $f(q, t_n) \in S(q, \delta)$ and, since $t_{2k-1} > 2k-1$ and $t_{2k} < -2k$, we have that $q \in A_q \cap \Omega_q$; consequently, the point q is Poisson stable. Since $q \in S$, this completes the proof of the theorem.

COROLLARY 1.14.   *The set of central motions is the closure of the set of Poisson stable points.*

Question 1.14.   Does there exist a Poisson stable point for a dynamical system defined in the plane R^2 and having at least one nonwandering point?

Question 2.14.   Does there exist a dynamical system (defined in some noncomplete space X) without wandering and Poisson stable points?

§ 15   Minimal center of attraction

In this section we shall consider only the values $t \in R^+$. Analogous concepts can also be defined for nonpositive t.

Suppose the set $E \subseteq X$ is closed or open. The *characteristic function* χ_E of the set E is defined by

$$\chi_E(p) = \begin{cases} 1 \text{ if } p \in E, \\ 0 \text{ if } p \in X \setminus E. \end{cases}$$

Consider the arc $f(p, [0, T])$ and the measurable set of those values of $t \in [0, T]$ for which $f(p, t) \in E$ (this set is measurable inasmuch as E is closed or open). The Lebesgue measure of this set is

$$\tau = \tau(p, T, E) = \int_0^T \chi_E(f(p, t))dt.$$

It is natural to call τ *the time of occurrence of the point p in the set E during the interval of time* $[0, T]$. It is clear that $0 \leq \tau/T \leq 1$. If

$$\lim_{T \to +\infty} \frac{1}{T} \int_0^T \chi_E(f(p, t))dt = \lim_{T \to +\infty} \frac{\tau}{T} \equiv P(f(p, t) \in E)$$

exists, then we call this limit the *probability of finding the point p in the set E.* Clearly (if these probabilities exist)

$$P(f(p, t) \in A \cup B) \leq P(f(p, t) \in A) + P(f(p, t) \in B),$$

and furthermore if $A \cap B = \phi$ then equality holds. Moreover, for $A \subseteq B \subseteq X$,

$$P(f(p, t) \in A) \leq P(f(p, t) \in B).$$

Definition 1.15.   An invariant closed set $V \subseteq X$ is a *center of attraction for* $f(p, t)$ if for any $\varepsilon > 0$

$$P(f(p, t) \in S(V, \varepsilon)) = 1. \tag{1.15}$$

If the set V does not contain a proper subset which is a center of attraction for $f(p, t)$ then V is called the *minimal* center of attraction for $f(p, t)$. (The theory of minimal centers of attraction was created by H. F. Hilmy [1].)

Denote by W_p the set consisting of all points $q \in X$ for which the inequality

$$\lim_{T \to +\infty} \sup \frac{\tau(p, T, S(q, \delta))}{T} > 0 \qquad (2.15)$$

is satisfied for every $\delta > 0$ (if such points exist). It is not difficult to see that

$$W_p \subseteq \Omega_p \subseteq \Sigma_p^+.$$

LEMMA 1.15. *The set W_p is invariant.*
Proof. Let $q \in W_p$, $t_0 \in R$, $\varepsilon > 0$. By condition 2.1, there exists a $\delta > 0$ such that

$$f(S(q, \delta), t_0) \subseteq S(f(q, t_0), \varepsilon).$$

Furthermore,

$$\tau(p, T, S(f(q, t_0), \varepsilon)) \geqq \tau(p, T, f(S(q, \delta), t_0)) \geqq \tau(p, T, S(q, \delta)) - |t_0|. \ (3.15)$$

It follows from (3.15) and (2.15) that

$$\lim_{T \to +\infty} \sup \frac{\tau(p, T, S(f(q, t_0), \varepsilon))}{T} > 0.$$

Since ε is arbitrary this means that $f(q, t_0) \in W_p$ and W_p is invariant.

LEMMA· 2.15. *The set W_p is closed.*
Proof. Let $q \in X \setminus W_p$. Then condition (2.15) is not satisfied for at least one $\delta = \delta_0 > 0$ and hence

$$P(f(p, t) \in S(q, \delta_0)) = 0. \qquad (4.15)$$

Let $q_1 \in S(q, \delta_0)$ and $\delta_1 > 0$ be such that

$$S(q_1, \delta_1) \subseteq S(q, \delta_0).$$

Then, in virtue of (4.15),

$$P(f(p, t) \in S(q_1, \delta_1)) = 0$$

and $q_1 \notin W_p$. Therefore

$$S(q, \delta_0) \subseteq X \setminus W_p$$

and the set $X \setminus W_p$ is open; consequently, W_p is closed. This completes the proof of the lemma.

LEMMA 3.15. *If V is a center of attraction for $f(p, t)$ then $W_p \subseteq V$.*
    Proof. Suppose $q \notin V$. Then

$$\alpha \equiv \rho(q, V) > 0, \, S\left(q, \frac{\alpha}{2}\right) \subseteq X \setminus S\left(V, \frac{\alpha}{2}\right).$$

It follows from the fact that V is a center of attraction for $f(p, t)$ that

$$P\left(f(p, t) \in S\left(V, \frac{\alpha}{2}\right)\right) = 1,$$

and then

$$P\left(f(p, t) \in X \setminus S\left(V, \frac{\alpha}{2}\right)\right) = 0$$

and *a fortiori*

$$P\left(f(p, t) \in S\left(q, \frac{\alpha}{2}\right)\right) = 0.$$

This equality means that

$$\lim_{T \to +\infty} \frac{\tau(p, T, S(q, \alpha/2))}{T} = 0,$$

and therefore $q \notin W_p$. It follows that $W_p \subseteq V$. This completes the proof of the lemma.

THEOREM 1.15. *If the motion $f(p, t)$ is positively Lagrange stable then the set $W_p \neq \phi$ and is the minimal center of attraction for $f(p, t)$.*

Proof. Suppose the motion is positively Lagrange stable. Then the set Σ_p^+ is compact. For arbitrary $\varepsilon > 0$ consider the compactum

$$K \equiv \Sigma_p^+ \setminus S(W_p, \varepsilon)$$

if $W_p \neq \phi$ and the compactum $K \equiv \Sigma_p^+$ if $W_p = \phi$. If $K \neq \phi$, then for any point $q \in K$ (inasmuch as $q \notin W_p$) there exists a $\delta = \delta_q > 0$ for which inequality (2.15) does not hold. Then

$$P(f(p, t) \in S(q, \delta_q)) = 0 \tag{5.15}$$

for all $q \in K$. The cover of K by the collection of neighborhoods $\{S(q, \delta_q)\}$ contains a finite subcover:

$$K \subseteq \bigcup_{i=1}^{m} S(q_i, \delta_{q_i}).$$

Furthermore, it follows from (5.15) that

$$P(f(p, t) \in K) = 0, \tag{6.15}$$

whereas obviously

$$P(f(p, t) \in \Sigma_p^+) = 1. \tag{7.15}$$

It follows from (6.15) and (7.15) that $W_p \neq \phi$ (otherwise these two equalities contradict one another) and

$$P(f(p, t) \in S(W_p, \varepsilon)) = 1. \tag{8.15}$$

But if $K = \phi$ then $S(W_p, \varepsilon) \supseteq \Sigma_p^+$ and, according to (7.15), equality (8.15) is again satisfied.

It follows, taking Lemmas 1.15 and 2.15 into consideration, that W_p is a center of attraction for $f(p, t)$. Minimality follows from Lemma 3.15. This completes the proof of the theorem.

THEOREM 2.15.   *If the space X is compact, then the minimal center of attraction W_p for $f(p, t)$ is contained in the set Z of central motions.*

Proof. On the basis of Remark 1.14 it suffices to prove that all the points in W_p are nonwandering in W_p. So, assume that there exists a point q in W_p which is wandering relative to W_p. Taking Remark 2.12 into consideration, we conclude that there exist an $\varepsilon > 0$ and a $t_0 > 0$ such that for all $t \geqq t_0$,

$$S_{W_p}(q, 2\varepsilon) \cap f(S_{W_p}(q, 2\varepsilon), t) = \phi,$$

where

$$S_{W_p}(q, 2\varepsilon) = S(q, 2\varepsilon) \cap W_p.$$

Then any point in $S_{W_p}(q, 2\varepsilon)$ must eventually abandon $S_{W_p}(q, 2\varepsilon)$; in view of the invariance of W_p, this must occur because the point abandons $S(q, 2\varepsilon)$ after time t_0.

Let $\eta > 0$ and $T_1 > t_0$ be such that

$$\frac{2t_0}{T_1} < \tfrac{1}{2}\eta. \tag{9.15}$$

For the compactum

$$K \equiv \bar{S}_{W_p}(q, \varepsilon)$$

and the numbers ε and T_1, from the condition of uniform integral continuity we find a corresponding $\delta > 0$, i.e., a $\delta > 0$ such that

$$\rho(f(r, t), f(r_0, t)) < \varepsilon$$

provided

$$r_0 \in K, \rho(r, r_0) < \delta, t \in [0, T_1].$$

We set

$$\Gamma = \bar{S}(K, \delta) \cap \bar{S}(q, \varepsilon), \Gamma' = \overline{S(q, \varepsilon) \setminus \Gamma}.$$

Let $T > T_1$ and consider the quantity $\tau(p, T, \Gamma)$. For every transversing of the point $p_1 = f(p, t)$ through Γ there exists a point $r_0 \in \bar{S}_{W_p}(q, \varepsilon)$ such that $\rho(p_1, r_0) < \delta$.

After the interval of time t_0 the point r_0 finally leaves $S(q, 2\varepsilon)$. According to the choice of δ,

$$\rho(f(p_1, t), f(r_0, t)) < \varepsilon$$

for all $t \in [0, T_1]$. Therefore, for $t \in [t_0, T_1]$, we have that

$$f(p_1, t) \notin S(q, \varepsilon).$$

Thus, after each stay of the point $f(p, t)$ in the domain Γ (the duration of each stay being less than or equal to t_0) it remains outside of $S(q, \varepsilon)$ during an interval of time $T_1 - t_0$. Then

$$\frac{\tau(p, T, \Gamma)}{T} \leqq \frac{2t_0}{T_1} . \qquad (10.15)$$

On the other hand, since $\Gamma' \cap W_p = \phi$, we have that

$$P(f(p, t) \in \Gamma') = 0.$$

Therefore, for sufficiently large T,

$$\frac{\tau(p, T, \Gamma')}{T} < \tfrac{1}{2}\eta. \qquad (11.15)$$

From (10.15), taking (9.15) and (11.15) into consideration, it follows that

$$\frac{\tau(p, T, S(q, \varepsilon))}{T} < \eta,$$

and, since $\eta > 0$ is arbitrary, that

$$P(f(p, t) \in S(q, \varepsilon)) = 0.$$

This contradicts the fact that $q \in W_p$. This completes the proof of the theorem.
    The following corollary is implied by (8.15) and Theorem 2.15:

COROLLARY 1.15 (G. D. BIRKHOFF [1]).   *If the space X is compact, then, for any point $p \in X$ and any $\varepsilon > 0$,*

$$P(f(p, t) \in S(Z, \varepsilon)) = 1.$$

As V. V. Stepanov [1] showed, if the function $\Phi(\varphi, \theta)$ in Example 2.10 satisfies the Lipschitz condition and

$$\iint_R \frac{d\varphi d\theta}{\Phi(\varphi, \theta)} = +\infty,$$

then, for any point $p \in X$ and any $\varepsilon > 0$,

$$P(f(p, t) \in S((0, 0), \varepsilon)) = 1.$$

Thus, in this case the rest point $(0, 0)$ is the minimal center of attraction for all motions. Here, at the same time, $Z = X$.

Chapter IV

MINIMAL SETS AND RECURRENT MOTIONS

§ 16   Minimal sets

Definition 1.16.   A *minimal set* is any nonempty closed invariant set which does not contain a proper nonempty closed invariant subset.

A rest point and also the trajectory of any periodic motion are compact minimal sets. The trajectory of a uniform linear motion is an example of a noncompact minimal set. Minimal sets were first investigated by G. D. Birkhoff [2].

THEOREM 1.16.   *Every closed invariant compact set F contains a minimal set.*

Proof. If the set F contains any proper closed invariant subset, then we denote this subset by F_1; otherwise we set $F_1 = F$. If F_1 contains a proper closed invariant subset, then we denote this subset by F_2; otherwise we set $F_2 = F_1$. Continuing this line of reasoning we obtain a nested sequence of closed invariant sets:

$$F \supseteq F_1 \supseteq F_2 \supseteq \ldots \supseteq F_k \supseteq \ldots.$$

Set

$$F_\omega \equiv \bigcap_{k=1}^{\infty} F_k.$$

Clearly, F_ω is a nonempty (since F is compact) closed invariant set. Being a closed subset of the compact set F, it is also compact, and therefore we can repeat the line of reasoning with respect to F_ω that we carried out for the set F, and so forth.

Thus a transfinite nested sequence of closed invariant sets is defined:

$$F \supseteq F_1 \supseteq \ldots \supseteq F_k \supseteq \ldots \supseteq F_\omega \supseteq \ldots \supseteq F_\alpha \supseteq \ldots \supseteq F_\beta \supseteq \ldots.$$

On the basis of the Baire-Hausdorff Theorem (as in § 14) we conclude that after at most a denumerable number of steps the sets in this sequence begin to coincide, i.e., there exists a γ such that $F_\gamma = F_{\gamma+1} = \ldots$. It is obvious that F_γ is a minimal set. This completes the proof of the theorem.

This theorem implies the following corollary.

COROLLARY 1.16. *If the space X is compact then it contains a minimal set.*
The following corollary also holds.

COROLLARY 2.16. *If the motion $f(p, t)$ is positively (negatively) Lagrange stable, then $\Omega_p(A_p)$ contains a minimal set.*

In fact, in this case $\Sigma_p^+(\Sigma_p^-)$ is compact and in virtue of inclusion (3.7) $\Omega_p(A_p)$ (as a closed invariant subset of a compact set) is a closed invariant compact set and consequently it contains a minimal set.

THEOREM 2.16 (CHARACTERIZING PROPERTY OF A MINIMAL SET). *A necessary and sufficient condition for an invariant set Σ to be minimal is that every trajectory $f(p, R)$ contained in Σ be everywhere dense in Σ, i.e., that $\Sigma_p = \Sigma$.*

Proof. Let Σ be a minimal set. Take an arbitrary point $p \in \Sigma$. Since Σ is invariant, $f(p, R) \subseteq \Sigma$, and, since Σ is closed, we have that $\Sigma_p \subseteq \Sigma$. But Σ_p cannot be a proper subset of the set Σ since in that case Σ would not be minimal – hence $\Sigma_p = \Sigma$. Suppose the equality $\Sigma_p = \Sigma$ holds for any trajectory belonging to the invariant set Σ. In this case Σ is also closed. It remains to prove that Σ does not contain a proper closed invariant subset. So suppose there exists a proper closed invariant subset $A \subset \Sigma$ and take an arbitrary point $p \in A$. Then $f(p, R) \subseteq A$ since A is invariant and $\Sigma_p \subseteq A$ since A is closed. Thus in this case

$$A \subset \Sigma = \Sigma_p \subseteq A.$$

The contradiction obtained proves the theorem.

THEOREM 3.16. *An isomorphism of dynamical systems carries a minimal set into a minimal set.*

The proof of this theorem follows from Corollary 3.6 if one takes into consideration that under an isomorphism of dynamical systems closed sets are carried into closed sets and a proper subset of a set is carried over into a proper subset of its image.

Note that under a homomorphism a minimal set may go over into a set which is not minimal. In fact, suppose X_1 consists of the rest point -1 and the interval $(0, 2\pi)$ on the real line R^1, which is the trajectory of a motion in the dynamical system $g(p, t)$. As the space X_2 take the unit circle. We stipulate an arbitrary point $x_0 \in X_2$ to be a rest point of the dynamical system $h(x, t)$. We map the interval $(0, 2\pi)$ onto $X_2 \setminus x_0$. The dynamical system $h(x, t)$ thus defined is a homomorphic image of the dynamical system $g(p, t)$. Furthermore, the interval $(0, 2\pi)$ in X_1 will be a minimal set whereas its image $X_2 \setminus x_0$ is not because it is not closed in X_2.

However, the following theorem holds.

THEOREM 4.16.   *A homomorphism of dynamical systems maps a compact minimal set into a compact minimal set.*

The proof of this theorem follows from Corollary 3.6 if one takes into consideration that a continuous mapping maps a compactum into a compactum and that the preimage of a proper subset of a set is a proper subset of the preimage of this set. [See, e.g., P. S. Aleksandrov [2], page 320, Theorem 8.] [By the preimage of a point $x \in X_2$ under the mapping ξ we understand the collection of all points $p \in X_1$ for which $\xi(p) = x$. The preimage of a set is understood to be the union of the preimages of all points of this set.]

§ 17   Almost recurrent motions and recurrent motions

Definition 1.17 (*V. V. Nemytsky* [2]).   We say that the trajectory $f(p, t)$ *uniformly approximates the set* $Q \subseteq X$ if for any $\varepsilon > 0$ there exists a number $T > 0$ such that every arc of the trajectory $f(p, R)$ of time length T approximates the set Q to within ε, i.e.,

$$Q \subseteq S(f(p, [t_0, t_0 + T]), \varepsilon) \tag{1.17}$$

for any t_0.

Definition 2.17.   A motion $f(p, t)$ is said to be *almost recurrent* if the trajectory $f(p, R)$ uniformly approximates the point p. (Almost recurrent motions were first studied by M. V. Bebutov [1, 2].)

THEOREM 1.17.   *Every almost recurrent motion is Poisson stable.*

Proof. Let $f(p, t)$ be an almost recurrent motion. Then for any $\varepsilon > 0$ there exists a $T > 0$ such that

$$p \in S(f(p, [t_0, t_0 + T]), \varepsilon) \tag{2.17}$$

for arbitrary t_0, i.e.,

$$S(p, \varepsilon) \cap f(p, [t_0, t_0 + T]) \neq \phi. \tag{3.17}$$

Since the t_0 in (3.17) is an arbitrary number, then by Theorem 1.7 the point $p \in \Omega_p$ and consequently it together with the motion $f(p, t)$ are Poisson stable.

Definition 3.17.   The motion $f(p, t)$ is said to be *recurrent* if the trajectory $f(p, R)$ uniformly approximates itself. (Recurrent motions were studied by G. D. Birkhoff [2].)

Clearly, every periodic motion with period τ is recurrent. In this case, for any $\varepsilon > 0$ we can take $T = \tau$. Then

$$f(p, [t_0, t_0 + T]) = f(p, R)$$

and the inclusion (1.17) holds for $Q = f(p, R)$.

Analogously, every rest motion is a recurrent motion. It is also obvious that the following theorem holds.

THEOREM 2.17.   *Every recurrent motion is almost recurrent and its trajectory is bounded.*

THEOREM 3.17.   *In a complete space every recurrent motion is Lagrange stable.*

Proof. Since the motion $f(p, t)$ is recurrent, then, for $\varepsilon/2 > 0$, there exists a number $T > 0$ such that $f(p, [0, T])$ approximates $f(p, R)$ to within $\varepsilon/2$, i.e.,

$$\rho(q, f(p, [0, T])) < \frac{\varepsilon}{2}$$

for all $q \in f(p, R)$. Then, for all $q \in \Sigma_p$,

$$\rho(q, f(p, [0, T])) \leqq \frac{\varepsilon}{2}. \tag{4.17}$$

But $f(p, [0, T])$ is a closed compact set (see Theorem 1.3) and therefore it possesses a finite $(\varepsilon/2)$-net which by inequality (4.17) is a finite (ε)-net for Σ_p. Then in virtue of the completeness of the space X the set Σ_p is compact, i.e., the motion $f(p, t)$ is Lagrange stable. [See L. A. Lyusternik and V. I. Sobolev [1], page 63, for the *Definition*: A set E of a metric space X is called an *ε-net* for the subset M of the same space if for any point $p \in M$ there exists a point $p_\varepsilon \in E$ such that $\rho(p, p_\varepsilon) < \varepsilon$; also see Theorem 3 (Hausdorff): For a closed set M in a metric space X to be compact it is necessary (and in the case X is complete also sufficient) that for any number $\varepsilon > 0$ there exist a finite ε-net for M.]

§ 18 Interrelationships among minimal sets, almost recurrent motions and recurrent motions

LEMMA 1.18.   *A motion $f(p, t)$ for which $\Omega_p \neq \phi$ spends, for $t > 0$, arbitrarily large intervals of time in any arbitrary neighborhood of an invariant set $A \subseteq \Omega_p$.*

Proof. Suppose the invariant set $A \subseteq \Omega_p$ and that $\varepsilon > 0$ and $T > 0$ are given numbers. Take an arbitrary point $q \in A$ and find a corresponding $\rho(q, \varepsilon, T) > 0$ from the condition for integral continuity (see Theorem 4.2) such that

$$\rho(f(r, t), f(q, t)) < \varepsilon$$

provided $\rho(r, q) < \delta$ and $0 \leqq t \leqq T$. Since the point $q \in \Omega_p$, there exists a point $p_1 \in f(p, R^+)$ in $S(q, \delta)$. Then $\rho(p_1, q) < \delta$ and by virtue of the choice of δ,

$$\rho(f(p_1, t), f(q, t)) < \varepsilon$$

for $0 \leqq t \leqq T$. It follows that

$$\rho(f(p_1, t), A) < \varepsilon$$

for $0 \leqq t \leqq T$, which is what was required to be proved.

THEOREM 1.18 (M. V. BEBUTOV [2]).   *The closure of the trajectory of an almost recurrent motion is a minimal set.*

Proof. Suppose the motion $f(p, t)$ is almost recurrent but that the set Σ_p is not minimal. Then there exists a proper closed invariant subset $A \subseteq \Sigma_p$. Clearly, the point $p \notin A$ since otherwise $\Sigma_p \subseteq A$ and $A = \Sigma_p$. Set $\rho(p, A) = d(d > 0)$ and take $\varepsilon < d/2(\varepsilon > 0)$. By virtue of the almost recurrence of the motion $f(p, t)$, for every ε there exists a $T > 0$ such that for any $t_0 \in R$ inclusion (2.17) holds. Moreover, by Theorem 1.17, $p \in \Omega_p$ and hence also $\Sigma_p \subseteq \Omega_p$. From this, taking (3.7) into consideration, we obtain the equality $\Omega_p = \Sigma_p$. On the basis of Lemma 1.18 there exists a segment $[t_0, t_0 + T]$ such that

$$f(p, [t_0, t_0 + T)) \subseteq S(A, \varepsilon). \tag{1.18}$$

According to (2.17) and (1.18), $p \in S(A, 2\varepsilon)$ and therefore

$$d = \rho(p, A) < 2\varepsilon < d.$$

The contradiction thus obtained proves the theorem.

THEOREM 2.18 (FIRST THEOREM OF BIRKHOFF [2]).   *A motion whose trajectory belongs to a compact minimal set is recurrent.*

Proof. Suppose $f(p, R)$ belongs to the compact minimal set Σ but that the motion $f(p, t)$ is not recurrent. Choose an arbitrary sequence of positive numbers $\{T_n\} \to +\infty$. Then there exists an $\varepsilon_0 > 0$ such that for any natural number n whatever there exists an arc $f(p_n, [-T_n, T_n])$ of the trajectory $f(p, R)$ of time length $2T_n$ which does not approximate $f(p, R)$ to within ε_0. Furthermore, there exists a point $q_n \in f(p, R)$ such that

$$q_n \notin S(f(p_n, [-T_n, T_n]), \varepsilon_0),$$

i.e.,

$$\rho(q_n, f(p_n, t)) \geq \varepsilon_0 \tag{2.18}$$

for all $t \in [-T_n, T_n]$.

We thus obtain two sequences of points $\{p_n\}$ and $\{q_n\}$ belonging to the trajectory $f(p, R)$ which are contained in the compact set Σ. These sequences contain convergent subsequences whose terms would have the same corresponding indices. In order not to complicate the notation let us assume that the sequences $\{p_n\}$ and $\{q_n\}$ themselves converge. Let

$$\lim_{n \to \infty} p_n = p^*, \lim_{n \to \infty} q_n = q^*.$$

Then $p^* \in \Sigma$ and $q^* \in \Sigma$ since these points belong to Σ_p and by the characterizing property of a minimal set $\Sigma_p = \Sigma$.

Let $t \in R$ be a fixed but arbitrary number. Then for sufficiently large n the number $t \in [-T_n, T_n]$ and inequality (2.18) is satisfied. Passing to the limit in it as $n \to \infty$ we obtain that

$$\rho(q^*, f(p^*, t)) \geqq \varepsilon_0.$$

Since this inequality holds for any $t \in R$,

$$\rho(q^*, f(p^*, R)) \geqq \varepsilon_0,$$

and consequently also $\rho(q^*, \Sigma_p^*) \geqq \varepsilon_0$. But $\Sigma_p^* = \Sigma$ since $p^* \in \Sigma$ and Σ is a minimal set. This means that $\rho(q^*, \Sigma) \geqq \varepsilon_0$, which however contradicts the fact that $q^* \in \Sigma$. This completes the proof of the theorem.

THEOREM 3.18 (SECOND THEOREM OF BIRKHOFF). *In a complete space the closure of the trajectory of a recurrent motion is a compact minimal set.*

This theorem obviously follows from Theorems 2.17, 1.18 and 3.17.

THEOREM 4.18 (M. V. BEBUTOV [2]). *A necessary and sufficient condition for a Lagrange stable motion to be recurrent is that it be almost recurrent.*

Proof. The *necessity* is obvious.

Sufficiency. Suppose that the motion $f(p, t)$ is Lagrange stable and almost recurrent. By Theorem 1.18, Σ_p is a minimal set. This set is compact by virtue of the Lagrange stability of $f(p, t)$. According to the first theorem of Birkhoff the motion $f(p, t)$ is recurrent.

§ 19 The Shcherbakov classification of Poisson stable motions. Pseudorecurrent motions

The fundamental results of this section are due to B. A. Shcherbakov [1–4].

Let the motion $f(p, t)$ be positively Poisson stable. This is equivalent to asserting that for any $\varepsilon > 0$, $t_0 \in R$ and $q \in f(p, R)$ there exists a function $T = T(\varepsilon, t_0, q) > 0$ such that

$$\rho(f(q, [t_0, t_0 + T]), q) < \varepsilon. \tag{1.19}$$

The last inequality can further be written in the following two equivalent forms:

$$q \in S(f(q, [t_0, t_0 + T]), \varepsilon), \tag{2.19}$$

$$f(q, [t_0, t_0 + T]) \cap S(q, \varepsilon) \neq \phi. \tag{3.19}$$

Note that in this connection the function $T(\varepsilon, t_0, q)$ is not defined uniquely if simply because of the fact that if the inequality (1.19) is satisfied for some function $T(\varepsilon, t_0, q)$ then it is also satisfied for any function

$$T_1(\varepsilon, t_0, q) \geq T(\varepsilon, t_0, q).$$

The following cases are possible:
1. The function T can be choosen so that it does not depend on ε. We shall show that this occurs if and only if the motion $f(p, t)$ is singular.

In fact, if the motion $f(p, t)$ is singular then there exists a number $\tau > 0$ such that $f(q, \tau) = q$ for arbitrary $q \in f(p, R)$. Then we can set $T = \tau$ in inequality (1.19) independently of ε since in this connection

$$f(q, [t_0, t_0 + \tau]) = f(p, R)$$

and

$$\rho(f(q, [t_0, t_0 + T]), q) = 0. \tag{4.19}$$

Now assume that the function T in inequality (1.19) does not depend on ε. Then equality (4.19) is satisfied for arbitrary $t_0 \in R$ and $q \in f(p, R)$. In the case $t_0 > 0$ and $q = p$ we obtain from (4.19) that

$$\rho(f(p, [t_0, t_0 + T]), p) = 0.$$

Therefore there exists a number $\tau \in [t_0, t_0 + T]$ ($\tau > 0$) such that $f(p, \tau) = p$ and consequently the motion $f(p, t)$ is singular.

2. The function T can be chosen so that it does not depend on t_0. This holds if and only if the motion $f(p, t)$ is almost recurrent. In fact, the function T in (2.19) can be selected so that it does not depend on t_0 if and only if the motion $f(q, t)$ is almost recurrent for any point $q \in f(p, R)$.

It now remains to prove that the following lemma holds.

LEMMA 1.19.   *If the motion* $f(p, t)$ *is almost recurrent, then the motion* $f(q, t)$ *is also almost recurrent for any point* $q \in f(p, R)$.

Proof. Let $q = f(p, t_1)$ and $\varepsilon > 0$. Since the function $f(p, t)$ is continuous in p there exists a $\delta > 0$ such that for $\rho(r, p) < \delta$ the inequality

$$\rho(f(r, t_1), q) < \varepsilon$$

is satisfied. For this δ, since the motion $f(p, t)$ is almost recurrent, there exists a corresponding number $T(\delta) > 0$, i.e., a $T > 0$, such that

$$\rho(f(p, [t_0, t_0 + T]), p) < \delta$$

for all t_0. Then, according to the choice of δ, inequality (1.19) with t_0 replaced by $t_0 + t_1$ is satisfied for all t_0, which is what was required to be proved.

3. The function T can be chosen so that it does not depend on the point q. Such motions were first investigated by B. A. Shcherbakov [1] – he called them *pseudorecurrent* motions.

THEOREM 1.19.   *Every pseudorecurrent motion is Poisson stable.*

Proof. That the motion is positively Poisson stable is clear. That it is negatively Poisson stable follows from the inequality

$$\rho(f(q, [-t_0 - T, -t_0]), q) \leqq \varepsilon, \tag{5.19}$$

which holds, together with inequality (1.19), for any $q \in f(p, R)$.

In fact, if inequality (1.19) holds for any point $q \in f(p, R)$, then on the segment

$$[t_0, t_0 + T(t_0, \varepsilon)]$$

there exists a number τ_0 such that

$$\rho(f(q, \tau_0), q) < \varepsilon,$$

a number τ_1 such that

$$\rho(f(q, -\tau_0+\tau_1), f(q, -\tau_0)) < \varepsilon,$$

..., a number τ_n such that

$$\rho(f(q, -\tau_{n-1}+\tau_n), f(q, -\tau_{n-1})) < \varepsilon, \tag{6.19}$$

and so forth. Since the sequence $\{\tau_n\}$ is bounded, it can be assumed to be convergent. Let $\{\tau_n\} \to \tau$. Then

$$\tau \in [t_0, t_0 + T],$$

and, passing to the limit as $n \to \infty$ in (6.19), we obtain

$$\rho(q, f(q, -\tau)) \leq \varepsilon,$$

from which inequality (5.19) follows.

4. The function T can be chosen so that it does not depend on t_0 and q. This holds if and only if the motion $f(p, t)$ is recurrent. In fact, if T in (2.19) does not depend on t_0 and q, then any point $q \in f(p, R)$ is contained in an ε-neighborhood of an arbitrary arc $f(q, [t_0, t_0 + T])$ (since t_0 is arbitrary) of the trajectory $f(p, R)$, i.e.,

$$f(p, R) \subseteq S(f(q, [t_0, t_0 + T]), \varepsilon) \tag{7.19}$$

for any $q \in f(p, R)$ and $t_0 \in R$, and therefore the motion $f(p, t)$ is recurrent. Conversely, if the motion $f(p, t)$ is recurrent, then there exists a $T = T(\varepsilon) > 0$ such that inclusion (7.19) holds for any $q \in f(p, R)$ and $t_0 \in R$, and then, in particular, inclusion (2.19) is satisfied for $T = T(\varepsilon)$ and any $q \in f(p, R)$ and $t_0 \in R$.

Clearly, *every recurrent motion is pseudorecurrent*.

THEOREM 2.19.   *All motions in the closure of a trajectory of a pseudorecurrent (recurrent) motion are pseudorecurrent (recurrent).*

The proof of this theorem follows from the fact that if one sets $T = T(\varepsilon/2, t_0)$ $[T = T(\varepsilon/2)]$ in inequality (1.19), then it will be satisfied not only for all $q \in f(p, R)$ but also for all points $q \in \Sigma_p$ since in this connection

$$\rho(f(q, [t_0, t_0 + T]), q) \leqq \frac{\varepsilon}{2} < \varepsilon.$$

COROLLARY 1.19.   *The closure of the trajectory of a recurrent motion is a minimal set which consists of recurrent trajectories* (see Theorems 2.17, 1.18 and 2.19).

THEOREM 3.19 (M. V. BEBUTOV [2]).   *All motions in the closure of the trajectory of an almost recurrent motion are almost recurrent.*

Proof. Assume that the motion $f(p, t)$ is almost recurrent and that the point $q \in \Sigma_p$. Let $\varepsilon > 0$ be arbitrary. Since $q \in \Sigma_p$, there exists a point $\bar{p} = f(p, \bar{t})$ such that

$$\rho(\bar{p}, q) < \frac{\varepsilon}{3}. \tag{8.19}$$

According to Lemma 1.19 the motion $f(\bar{p}, t)$ is almost recurrent. Therefore there exists a $T(\varepsilon/3) > 0$ such that

$$\rho(\bar{p}, f(\bar{p}, [t_0, t_0 + T])) < \frac{\varepsilon}{3} \tag{9.19}$$

for any $t_0 \in R$.

Let $t' \in R$. For the point $f(q, t')$ and the numbers $\varepsilon/3$ and $T(\varepsilon/3)$ there exists by the integral continuity condition a $\delta > 0$ such that

$$\rho(f(r, t), f(q, t' + t)) < \frac{\varepsilon}{3}$$

for

$$\rho(r, f(q, t')) < \delta$$

and any $t \in [0, T]$. Since $\Sigma_{\bar{p}} = \Sigma_p$ and $q \in \Sigma_p$, then $f(q, t') \in \Sigma_{\bar{p}}$ and therefore there exists a t_0 such that

$$\rho(f(q, t'), f(\bar{p}, t_0)) < \delta.$$

For this t_0, by virtue of (9.19), there exists a $\tau \in [0, T]$ such that

$$\rho(\bar{p}, f(\bar{p}, t_0+\tau)) < \frac{\varepsilon}{3}. \tag{10.19}$$

Moreover, by virtue of the choice of the number δ,

$$\rho(f(q, t'+\tau), f(\bar{p}, t_0+\tau)) < \frac{\varepsilon}{3}. \tag{11.19}$$

From (11.19), (10.19) and (8.19) we obtain that

$$\rho(f(q, t'+\tau), q) < \varepsilon.$$

Thus

$$\rho(f(q, [t', t'+T]), q) < \varepsilon$$

for any $t' \in R$ and so the motion $f(q, t)$ is almost recurrent. This completes the proof of the theorem.

Now the theorem just proved and Theorem 1.18 imply the following corollary.

COROLLARY 2.19.   *The closure of the trajectory of an almost recurrent motion is a minimal set consisting of almost recurrent trajectories.*

THEOREM 4.19.   *A necessary and sufficient condition for a Lagrange stable motion $f(p, t)$ to be pseudorecurrent is that all motions in Σ_p be positively Poisson stable.*

Proof. Necessity follows from Theorems 2.19 and 1.19. To prove the sufficiency we suppose that the Lagrange stable motion $f(p, t)$ is not pseudo-recurrent and choose an arbitrary sequence of positive numbers $\{T_n\} \to +\infty$. Then there exist $\varepsilon > 0$ and $t_0 \in R$ such that for any natural number n whatever there exists a point $q_n \in f(p, R)$ such that

$$\rho(f(q_n, [t_0, t_0+T_n]), q_n) \geqq \varepsilon. \tag{12.19}$$

In view of the Lagrange stability of the motion $f(p, t)$, the sequence $\{q_n\}$ can be assumed to be convergent. Let $\{q_n\} \to q(q \in \Sigma_p)$. Then it follows from (12.19) that

79

$$\rho(f(q, [t_0, +\infty)), q) \geqq \varepsilon.$$

Thus the point q cannot be positively Poisson stable.

A theorem analogous to Theorem 4.19 can be proved if in the latter one replaces positive Poisson stability by negative Poisson stability.

THEOREM 5.19.   *A homomorphism of dynamical systems with compact (complete) phase space X_1 carries pseudorecurrent (recurrent) motions into pseudorecurrent (recurrent) motions.*

Proof. To prove this in the case of a pseudorecurrent motion one must use Theorem 4.19, Lemma 1.9, and the fact that a homomorphism carries the closure of a set which is compact in X_1 into the closure of its image which is compact in X_2. In the case of a recurrent motion it suffices to use Theorems 3.18, 4.16 and 2.18.

THEOREM 6.19.   *A homomorphism of dynamical systems carries almost recurrent motions into almost recurrent motions.*

Proof. Let $\xi: X_1 \to X_2$ be a continuous mapping which realizes a homomorphism of the dynamical system $g(p, t)$ into the dynamical system $h(x, t)$. Assume that the motion $g(p, t)$ is almost recurrent, $x = \xi(p)$, and $\varepsilon > 0$. By virtue of the continuity of the mapping ξ, there exists an $\eta > 0$ such that for $\rho(q, p) < \eta$ $(p, q \in X_1)$ the inequality

$$\rho(\xi(q), \xi(p)) < \varepsilon \tag{13.19}$$

is satisfied. From the condition that the motion $g(p, t)$ is an almost recurrent motion, for the number η there exists a $T > 0$ such that

$$p \in S(g(p, [t_0, t_0 + T]), \eta)$$

for any $t_0 \in R$. Then, in accordance with the choice of η,

$$\xi(p) \in S(\xi(g(p, [t_0, t_0 + T])), \varepsilon).$$

From this, taking Lemma 1.6 into consideration, we obtain that

$$x \in S(h(x, [t_0, t_0 + T]), \varepsilon),$$

i.e., the motion $h(x, t)$ is almost recurrent.

§ 20 The Bebutov dynamical system

It is convenient to construct examples of various types of motions in the so-called *Bebutov dynamical system* (see Bebutov [1, 2]).

Bebutov space X_u consists of the set of all continuous functions φ on $(-\infty, +\infty)$ with metric ρ defined by

$$\rho(\varphi_1, \varphi_2) = \sup_{x_0 > 0} \min \{ \max_{|x| \le x_0} |\varphi_2(x) - \varphi_1(x)|; \ 1/x_0 \}.$$

LEMMA 1.20. *The inequality*

$$\rho(\varphi_1, \varphi_2) \le \varepsilon \tag{1.20}$$

holds if and only if

$$\max_{|x| \le 1/\varepsilon} |\varphi_2(x) - \varphi_1(x)| \le \varepsilon. \tag{2.20}$$

Proof. Suppose inequality (1.20) is satisfied and assume that

$$|\varphi_2(x) - \varphi_1(x)| > \varepsilon \quad \text{and} \quad |x| \le 1/\varepsilon.$$

Since $\varphi_2 - \varphi_1$ is continuous we can assume that $|x| < 1/\varepsilon$. Then

$$\min \{ \max_{|x| \le |x|} |\varphi_2(x) - \varphi_1(x)|; \ 1/|x| \} > \varepsilon$$

and *a fortiori* that $\rho(\varphi_1, \varphi_2) > \varepsilon$, which contradicts inequality (1.20).

Now assume that inequality (2.20) is satisfied. Take an arbitrary $x_0 > 0$. Then, if $x_0 \ge 1/\varepsilon$, we have that

$$\min \{ \max_{|x| \le x_0} |\varphi_2(x) - \varphi_1(x)|; \ 1/x_0 \} \le 1/x_0 \le \varepsilon,$$

and for $x_0 \le 1/\varepsilon$ we have that

$$\min \{ \max_{|x| \le x_0} |\varphi_2(x) - \varphi_1(x)|; \ 1/x_0 \} \le \max_{|x| \le 1/\varepsilon} |\varphi_2(x) - \varphi_1(x)| \le \varepsilon$$

(taking inequality (2.20) into consideration). Thus inequality (1.20) holds. This completes the proof of the lemma.

The following corollary is easily established on the basis of Lemma 1.20, just proved.

COROLLARY 1.20. $\rho(\varphi_n, \varphi) \to 0$ *if and only if* $\{\varphi_n(x)\}$ *converges to* $\varphi(x)$ *uniformly on every finite interval.*

[A space X is called *separable* if and only if there exists a countable everywhere dense set in X. The concept of separability for a space is equivalent to the concept of a countable basis (see P. S. Aleksandrov [2], p. 270, Theorem 47).]

It is not difficult to demonstrate the completeness and separability of the space X_u. For example, the set of all polynomials with rational coefficients forms a countable everywhere dense set in it.

Define a dynamical system $f_u(\varphi, t)$ in X_u by setting

$$f_u(\varphi, t) = \psi,$$

where

$$\psi(x) \equiv \varphi(x+t).$$

The dynamical system constructed in this way is called the *Bebutov dynamical system* (or the *shift dynamical system*).

Example 1.20.   Let us construct a two-sided sequence

$$\ldots, a_{-k}, \ldots, a_{-2}, a_{-1}, a_0, a_1, a_2, \ldots, a_k, \ldots \tag{3.20}$$

in the following way. We set a_i equal to zero for all even subscripts:

$$a_{2m} = 0 \quad \text{for} \quad m = 0, \pm 1, \pm 2, \ldots. \tag{3.20}$$

The remaining a_i (i.e., those with odd subscripts) are assigned values as follows. Every odd number $2m+1$ can be uniquely represented in the form $2m+1 = 3^{n-1}(6s \pm 1)$, where n is a natural number and s is an integer. Set $a_{2m+1} = n$.

Now setting $\varphi(k) = a_k$ for all integers k we extend the definition of the function $\varphi(x)$ to the entire axis $(-\infty, +\infty)$ by linear interpolation. It is not difficult to show that the motion

$$f_u(\varphi, t) = \varphi(x+t),$$

which is defined by the function constructed above in the Bebutov dynamical system, is *almost recurrent*. In fact, the segment

$$[n = a_{-3n-1}, \ldots, a_{-2}, a_{-1}, a_0, a_1, a_2, \ldots, a_{3n-1} = n]$$

in (3.20) repeats an infinite number of times to the left and right over intervals equal to $2 \cdot 3^n$. In fact, if $-3^{n-1} \leq m \leq 3^{n-1}$ and m is odd, then $m = 3^{n'-1}(6s \pm 1)$, where $n' \leq n$. Then for arbitrary integer σ the number $m + 2 \cdot 3^n \cdot \sigma = 3^{n'-1}(6s \pm 1 + 6\sigma')$ and therefore $a_{m + 2 \cdot 3^n \cdot \sigma} = a_m = n'$. This guarantees the almost recurrence of $f_u(\varphi, t)$.

We now consider the problem of determining, by means of the properties of the function $\varphi \in X_u$, the trajectory $f_u(\varphi, R)$ passing through it.

THEOREM 1.20.   *A necessary and sufficient condition for the function $\varphi \in X_u$ to be a rest point in the Bebutov dynamical system is that $\varphi(x) = $ constant.*

THEOREM 2.20.   *A necessary and sufficient condition for the motion $f_u(\varphi, t)$ in the Bebutov dynamical system to be periodic with period τ is that the function $\varphi(x)$ be periodic with the same period.*

We shall now prove that the following theorem holds.

THEOREM 3.20.   *A necessary and sufficient condition for the motion $f_u(\varphi, t)$ in the Bebutov dynamical system to be Lagrange stable is that the function $\varphi(x)$ be bounded and uniformly continuous on the entire real line.*

Proof. Suppose the motion $f_u(\varphi, t)$ is Lagrange stable. Consider the following family of functions:

$$\{\varphi(x \pm n)\} \quad (n = 0, 1, 2, \ldots). \tag{4.20}$$

All of them belong to the trajectory $f_u(\varphi, t)$. Since the motion $f_u(\varphi, t)$ is Lagrange stable the set Σ_φ is compact. Therefore every sequence of functions from the family (4.20) contains a subsequence which converges in X_u, and which according to Corollary 1.20 will in particular be uniformly convergent on $[0, 1]$. By Arzelà's theorem, the family of functions (4.20) is uniformly bounded and equicontinuous on $[0, 1]$. It follows from this that the function $\varphi(x)$ is bounded and uniformly continuous on the entire real line.

Assume now that the function $\varphi(x)$ is bounded and uniformly continuous on the entire real line. Take an arbitrary sequence $\{f_u(\varphi, t_n)\}$, i.e., $\{\varphi(x+t_n)\}$. It is uniformly bounded and equicontinuous on the entire real line. Making use of Arzela's theorem, select a subsequence $\{\varphi(x+t_n^1)\}$ from it which converges uniformly on $[-1, 1]$. In an analogous way, select from this sequence a subsequence $\{\varphi(x+t_n^2)\}$ which converges uniformly on $[-2, 2]$, and so forth. Then the diagonal sequence $\{\varphi(x+t_n^n)\}$ will be uniformly convergent on any finite interval, i.e., $\{f_u(\varphi, t_n^n)\}$ will converge in X_u. Consequently, $f_u(\varphi, R)$ is sequentially compact in X_u, and so the motion $f_u(\varphi, t)$ is Lagrange stable. This completes the proof of the theorem.

The function $\varphi(x)$ in Example 1.20 is unbounded and therefore the motion $f_u(\varphi, t)$ is not Lagrange stable. Then, according to Theorem 3.17, it cannot be recurrent (although it is almost recurrent).

The Bebutov dynamical system is "universal" in a certain sense since dynamical systems of rather extensive classes can be mapped isomorphically into it. For instance, the following theorem holds.

THEOREM 4.20.   *A necessary and sufficient condition for a dynamical system $f(p, t)$, defined on a compact space X, to be isomorphic to some subsystem of the Bebutov dynamical system is that its set of rest points be homeomorphic to some subset of the real line R.*

Proof. Necessity follows from Theorem 1.20 since the set of constants in Bebutov space is homeomorphic to the real line R. This was observed by B. A. Shcherbakov [7]. To prove the *sufficiency*, which was established by M. V. Bebutov [1, 2] for the case of a system with one rest point and by S. Kakutani [1] for the general case, assume that X is a compact space and that M is the set of rest points of the dynamical system $f(p, t)$ defined in X, M is homeomorphic to a subset F of the real line, and

$$t = \gamma(p)(p \in M, t \in F)$$

is a function which realizes this homeomorphism. Let $\Gamma(p)$ denote an arbitrary continuous real-valued function on X which coincides with $\gamma(p)$ on M and denote the set of all such functions $\Gamma(p)$ by Φ. Then by the extension theorem the set $\Phi \neq \phi$. [See P. S. Aleksandrov [2], p. 284, Theorem 66: Given any bounded continuous function γ on a closed set M of a metric space X, there exists a function which is continuous on the whole space X and coincides with γ at all points of the set M. Furthermore, if μ_0 is the least upper bound of the functions $|\gamma|$ on M, then the function Γ can be chosen so that the least upper bound of its absolute value (over the entire space X) is also the number μ_0.] If we set

$$\|\Gamma\| = \sup_{p \in X} |\Gamma(p)|, \ \rho(\Gamma_1, \Gamma_2) = \|\Gamma_1 - \Gamma_2\|,$$

then Φ will be a complete separable metric space.

Denote the direct products $X \times X$ and $M \times M$ by X^* and M^*, respectively, and set

$$\Delta^* = \{(p, p): p \in X\}, \ D^* = X^* \setminus (M^* \cup \Delta^*).$$

We shall say that the function $\Gamma \in \Phi$ decomposes *the motion on the set* $U^* \subseteq X^*$ if for an arbitrary point $(p, q) \in U^*$ there exists a $t \in R$ such that

$$\Gamma(f(p, t)) \neq \Gamma(f(q, t)).$$

We denote the collection of all functions which decompose motions on the set U^* by $\Phi(U^*)$.

We shall first prove three lemmas.

LEMMA 2.20.   *For every set $U^* \subseteq X^*$ the set $\Phi(\bar{U}^*)$ is open in Φ.*

Proof. Let $U^* \subseteq X^*$ and $\Gamma_0 \in \Phi(\bar{U}^*)$. Define a metric d by

$$d(p, q) = \sup_{t \in R} |\Gamma_0(f(p, t)) - \Gamma_0(f(q, t))|.$$

Then for any point $(p, q) \in \bar{U}^*$ we have that $d(p, q) > 0$. We shall show that there exists a number $\delta_0 > 0$ such that $d(p, q) \geq \delta_0$ for all points $(p, q) \in \bar{U}^*$. For suppose the contrary, i.e., assume that there exists a sequence $\{(p_n, q_n)\}$ of points in $\bar{U}^*$ such that $\{d(p_n, q_n)\} \to 0$. In view of the fact that X is compact we may assume that the sequence $\{(p_n, q_n)\}$ converges. Let $\{(p_n, q_n)\} \to (p_0, q_0)$. Inasmuch as for any $t' \in R$,

$$|\Gamma_0(f(p_n, t)) - \Gamma_0(f(q_n, t))| \leq d(p_n, q_n),$$

in the limit we obtain that for every $t \in R$

$$\Gamma_0(f(p_0, t)) = \Gamma_0(f(q_0, t)),$$

and this contradicts the fact that $(p_0, q_0) \in \bar{U}^*$ and $\Gamma_0 \in \Phi(\bar{U}^*)$.

It is easy to see that if we take $S(\Gamma_0, \delta_0/2)$ in Φ, then

$$S\left(\Gamma_0, \frac{\delta_0}{2}\right) \subseteq \Phi(\bar{U}^*),$$

and therefore $\Phi(\bar{U}^*)$ is open in Φ.

LEMMA 3.20.   *For any point (p_0, q_0) in D^* there exists a neighborhood U^* of it such that $\Phi(\bar{U}^*)$ is dense in Φ.*

Proof. Since $(p_0, q_0) \notin M^* \cup \Delta^*$, we can assume that $p_0 \notin M$ and $q_0 \neq p_0$. Let U denote a neighborhood of the point p_0 such that

$$\bar{U} \cap (M \cup q_0) = \phi.$$

Since p_0 is not a rest point there exists an $\eta > 0$ such that $f(p_0, \eta) \neq p_0$ and

$$f(p_0, [0, \eta]) \cap f(X \setminus U, [0, \eta]) = \phi.$$

Choose a neighborhood W of the point p_0 and a number $\delta \in (0, \eta)$ for which the sets

$$A \equiv f(\bar{W}, [0, \delta]),$$

$$B \equiv f(\bar{W}, [\eta, \eta + \delta]),$$

$$C \equiv f(X \setminus U, [0, \eta + \delta])$$

are pairwise disjoint.

Set

$$u(p) = \frac{\rho(p, A \cup C)}{\rho(p, A \cup C) + \rho(p, B)}.$$

Then $u(p)$ is a continuous function on X,

$$0 \leq u(p) \leq 1 \text{ in } X,$$

$u(p) = 0$ on $A \cup C$, and

$u(p) = 1$ on B.

Since $M \subseteq X \setminus U \subseteq C$, $u(p) = 0$ on the set M.
Define the function $v(p)$ by

$$v(p) = \int_0^\eta u(f(p, s)) \mathrm{d}s \quad (p \in X).$$

It is easy to see that the function $v(p)$ is continuous on X and that

$$0 \leq v(p) \leq \eta \text{ in } X,$$

$$v(f(p, t)) = v(p) + t \quad \text{if} \quad p \in \bar{W}, 0 \leq t \leq \delta, \tag{5.20}$$

$$v(f(p, t)) = 0 \quad \text{if} \quad p \in X \setminus U, 0 \leq t \leq \delta, \tag{6.20}$$

$$v(f(p, t)) = 0 \quad \text{if} \quad p \in M, t \in R.$$

We shall verify equality (5.20). Clearly,

$$v(f(p, t)) = \int_0^\eta u(f(p, t+s)) \mathrm{d}s = \int_t^{t+\eta} u(f(p, s)) \mathrm{d}s$$

$$= \int_0^\eta u(f(p, s)) \mathrm{d}s + \int_\eta^{\eta+t} u(f(p, s)) \mathrm{d}s - \int_0^t u(f(p, s)) \mathrm{d}s.$$

The first term in the extreme right member is $v(p)$, the second term equals t since $u(p) = 1$ on B, and the third term equals 0 in view of the fact that $u(p) = 0$ on A. Therefore (5.20) holds.

Finally, set

$$U^* = W \times (X \setminus \bar{U}).$$

It is clear that U^* is a neighborhood of the point (p_0, q_0) in D^* and that $\bar{U}^* = \bar{W} \times (X \setminus U)$. We shall show that $\Phi(\bar{U}^*)$ is dense in Φ.

Let $\Gamma \in \Phi$ and $\varepsilon > 0$. Setting

$$\Gamma_\alpha(p) = \frac{1}{\alpha} \int_0^\alpha \Gamma(f(p, s)) ds,$$

we can find an $\alpha > 0$ such that

$$\rho(\Gamma, \Gamma_\alpha) < \frac{\varepsilon}{2}. \tag{7.20}$$

Clearly, $\Gamma_\alpha(p) = \Gamma(p) = \gamma(p)$ on M and therefore $\Gamma_\alpha \in \Phi$. Since

$$\frac{\partial}{\partial t} \Gamma_\alpha(f(p, t)) = \frac{\partial}{\partial t} \left(\frac{1}{\alpha} \int_t^{t+\alpha} \Gamma(f(p, s)) ds \right) = \frac{1}{\alpha}[\Gamma(f(p, t+\alpha)) - \Gamma(f(p, t))],$$

we have that

$$\left| \frac{\partial}{\partial t} \Gamma_\alpha(f(p, t)) \right| \leq \frac{2\|\Gamma\|}{\alpha} \tag{8.20}$$

for arbitrary $t \in R$.

Now choose (positive) integers n and m such that

$$m > \frac{2}{\varepsilon}, \tag{9.20}$$

$$n > \max\left(\frac{4m\|\Gamma\|}{\alpha}, \frac{\pi}{\delta} \right) \tag{10.20}$$

and set

$$g = \Gamma_\alpha + \frac{1}{m} \sin nv.$$

Clearly $g \in \Phi$ since $v|_M = 0$. Furthermore, according to (7.20) and (9.20),

$$\|\Gamma - g\| \leq \|\Gamma - \Gamma_\alpha\| + \frac{1}{m} < \varepsilon.$$

It remains to show that $g \in \Phi(\bar{U}^*)$.

Obviously,

$$\left| \frac{\partial}{\partial t} g(f(p, t)) \right| \geqq \left| \frac{\partial}{\partial t} \frac{1}{m} \sin \left[nv(f(p, t)) \right] \right| - \left| \frac{\partial}{\partial t} \Gamma_\alpha(f(p, t)) \right|$$

(11.20)

$$\geqq \left| \frac{n}{m} \cos \left[nv(f(p, t)) \right] \frac{\partial}{\partial t} v(f(p, t)) \right| - \frac{2\|\Gamma\|}{\alpha}.$$

Since, according to (10.20), $\pi/n < \delta$, then, in view of (5.20), for arbitrary $p \in \bar{W}$ there exists at least one $t \in (0, \delta)$ such that

$$nv(f(p, t)) \equiv 0 \pmod{\pi}.$$

For such a p and t, we obtain from (11.20), taking (10.20) into consideration, that

$$\left| \frac{\partial}{\partial t} g(f(p, t)) \right| \geqq \frac{n}{m} - \frac{2\|\Gamma\|}{\alpha} > \frac{2\|\Gamma\|}{\alpha}.$$

(12.20)

On the other hand, according to (6.20), for arbitrary $q \in X \setminus U$ and $0 < t < \delta$,

$$g(f(q, t)) = \Gamma_\alpha(f(q, t)).$$

From this equality and (8.20) it follows that

$$\left| \frac{\partial}{\partial t} g(f(q, t)) \right| \leqq \frac{2\|\Gamma\|}{\alpha}$$

(13.20)

for arbitrary $q \in X \setminus U$ and $0 < t < \delta$.

Now (12.20) and (13.20) imply that for any $p \in \bar{W}$ and $q \in X \setminus U$ there exists a $t \in (0, \delta)$ such that

$$\frac{\partial}{\partial t} g(f(p, t)) \neq \frac{\partial}{\partial t} g(f(q, t)),$$

and therefore $g \in \Phi(\bar{U}^*)$, which is what was to be proved.

LEMMA 4.20.   *If the space X is compact, the set M of rest points of the*

89

*dynamical system $f(p, t)$ is homeomorphic to a subset of the real line R, and
γ is the continuous function defined on M which realizes this homeomorphism,
then the family of functions $\Phi(X^* \setminus \Delta^*)$ determined by γ is nonempty.*

Proof. By Lemma 3.20, there exists for each point $(p_0, q_0) \in D^*$ a
neighborhood U^* in D^* such that $\Phi(\bar{U}^*)$ is dense in Φ. The union of all
these neighborhoods is D^*. Since the set D^* is a separable metric space,
the above collection of neighborhoods contains a sequence of neighborhoods
$\{U_n^*\}$ such that

$$\bigcup_{n=1}^{\infty} U_n^* = D^*.$$

[See P. S. Aleksandrov [2], page 273, Theorem 52: An arbitrary uncountable
collection of open sets in a separable space contains a denumerable or finite
subcollection having the same union as the entire collection.]

Since by Lemma 2.20 every $\Phi(\bar{U}_n^*)$ is open in Φ (and Φ is complete),

$$\bigcap_{n=1}^{\infty} \Phi(\bar{U}_n^*) \neq \phi.$$

[See P. S. Aleksandrov [2], page 363, Theorem 27: If each set in a countable
collection of open sets in a complete metric space X is dense in X, then the
intersection of this collection is also dense in X.]

It is easy to see that

$$\bigcap_{n=1}^{\infty} \Phi(\bar{U}_n^*) \subseteq \Phi(X^* \setminus \Delta^*),$$

and therefore

$$\Phi(X^* \setminus \Delta^*) \neq \phi.$$

Proof of sufficiency in Theorem 4.20.   Let X be compact, let the set
M of rest points of the dynamical system $f(p, t)$ be homeomorphic to a
subset of the real line R, and let γ be a continuous real-valued function
defined on M which realizes this homeomorphism. By Lemma 4.20, there
exists a function $\Gamma \in \Phi(X^* \setminus \Delta^*)$.

Denote by σ the mapping $\varphi = \sigma(p)$ of the space X into the Bebutov space X_u such that

$$\varphi(t) = \Gamma(f(p, t)).$$

It is easy to see that σ is a 1–1 and continuous mapping of X into X_u. Since X is compact, σ is a homeomorphism of X onto the subset $\sigma(X)$ of the space X_u.

Setting

$$\psi = \sigma(f(p, \tau)),$$

we have, according to the definition of σ, that

$$\psi(t) = \Gamma(f(p, t+\tau)),$$

and therefore

$$\psi = f_u(\varphi, \tau),$$

i.e.,

$$\sigma(f(p, \tau)) = f_u(\sigma(p), \tau).$$

Thus σ realizes an isomorphism of the dynamical system $f(p, t)$ into a subsystem of the Bebutov dynamical system.

Chapter V

ALMOST PERIODIC MOTIONS. LYAPUNOV STABILITY

§ 21   Uniformly Poisson stable motions and almost periodic motions

Suppose the motion $f(p, t)$ is positively Poisson stable. It follows from inequality (1.19) that there exists at least one function $\tau = \tau(\varepsilon, t_0, q)$ such that

$$t_0 \leqq \tau(\varepsilon, t_0, q) \leqq t_0 + T(\varepsilon, t_0, q) \tag{1.21}$$

and

$$\rho(f(q, \tau), q) < \varepsilon. \tag{2.21}$$

A number τ which satisfies inequality (2.21) is said to be an ε-*displacement of the point q*.

If the function τ can be chosen so that it does not depend on the point $q \in f(p, R)$, then the motion $f(p, t)$ is called *uniformly Poisson stable*. In the case when the inequality (2.21) is satisfied for all points q in some set $A \subseteq X$ we shall say that τ is an ε-displacement of the set A. It is clear that we can make the following definition.

Definition 1.21.   A motion $f(p, t)$ is called *uniformly Poisson stable* if for any $\varepsilon > 0$ and $t_0 \in R$ there exists an ε-displacement $\tau(\varepsilon, t_0) > t_0$ of the trajectory $f(p, R)$. [In this case the number $-\tau(\varepsilon, t_0)$ is also an ε-displacement of the trajectory. In fact, if (2.21) holds for all points $q \in f(p, R)$, then, replacing q by $f(q, -\tau)$ in it, we arrive at the inequality $\rho(q, f(q, -\tau)) < \varepsilon$.] [Uniformly Poisson stable motions were first considered by M. V. Bebutov [1, 2].]

THEOREM 1.21.   *Every uniformly Poisson stable motion is pseudorecurrent.*

In fact, if in (1.21) τ does not depend on q, then we may also choose T to be independent of q in these inequalities. It suffices to set $T = \tau - t_0$.

COROLLARY 1.21.   *Every uniformly Poisson stable motion is Poisson stable.*

If the functions τ and T in inequalities (1.21) can be chosen so that τ does not depend on q and T does not depend on t_0 and q, then the motion $f(p, t)$ is said to be almost periodic.

Obviously, this definition is equivalent to the following one.

Definition 2.21.   The motion $f(p, t)$ is *almost periodic* if for any $\varepsilon > 0$ there exists a $T(\varepsilon) > 0$ such that in any interval of time of length T there exists a number τ such that for all $t \in R$ the inequality

$$\rho(f(p, t+\tau), f(p, t)) < \varepsilon \tag{3.21}$$

holds.

In other words, a motion $f(p, t)$ is almost periodic if for any $\varepsilon > 0$ there exists a $T > 0$ such that on any time segment $[t_0, t_0 + T]$ of length T there exists at least one ε-displacement τ of an arbitrary point $q = f(p, t)$ of the trajectory $f(p, R)$. Almost periodic motions were studied by P. Franklin [1].

THEOREM 2.21.   *Every singular motion is almost periodic.*

We shall prove the theorem for the case of a periodic motion $f(p, t)$ with period τ_0. If for any $\varepsilon > 0$ we take $T = \tau_0$ then the numbers $\tau = k\tau_0$, with k an integer, will be ε- (and even 0-) displacements of any point $q \in f(p, R)$ and furthermore it is obvious that there exists a number of the form $k t_0$ on any segment $[t_0, t_0 + T]$ of length T.

THEOREM 3.21.   *Every almost periodic motion is recurrent and uniformly Poisson stable.*

This theorem follows directly from the definition ($\tau = \tau(\varepsilon, t_0)$, $T = T(\varepsilon)$).

The converse assertion is false. B. A. Shcherbakov showed in [2] that there exists a uniformly Poisson stable recurrent motion which is not almost periodic. Here, the situation is that inequalities (1.21) have the form

$$t_0 \leqq \tau(\varepsilon, t_0) \leqq t_0 + T(\varepsilon, t_0) \tag{4.21}$$

for a uniformly Poisson stable motion,

$$t_0 \leqq \tau_1(\varepsilon, t_0, q) \leqq t_0 + T_1(\varepsilon) \tag{5.21}$$

for a recurrent motion, and

$$t_0 \leqq \tau_2(\varepsilon, t_0) \leqq t_0 + T_2(\varepsilon) \tag{6.21}$$

for an almost periodic motion.

[Note that if in (1.21) τ does not depend on ε, then the motion $f(p, t)$ is singular. τ cannot be independent of t_0 if only because $\tau \geqq t_0$.]

THEOREM 4.21.   *In the closures of trajectories of an almost periodic (uniformly Poisson stable) motion, all motions are almost periodic (uniformly Poisson stable).*

Remark 1.21.   The proof follows from the fact that for $\tau = \tau(\varepsilon/3, t_0)$ inequality (2.21) is satisfied for all $q \in \Sigma_p$ as well as for all $q \in f(p, R)$.

Definition 3.21.   A set $\{\tau\}$ of real numbers τ is *relatively dense* if there exists an $L > 0$ such that at least one point of the set $\{\tau\}$ is in any segment $[l, l+L] \subset R$.

Remark 2.21.   It is easy to see that a motion $f(p, t)$ is almost periodic (uniformly Poisson stable) if for any $\varepsilon > 0$ there exists a relatively dense (unbounded) set of ε-displacements of the trajectory $f(p, R)$.

Definition 4.21.   A function $\varphi(x)$ defined on R is *Bohr almost periodic* (see H. Bohr [1]) if for any $\varepsilon > 0$ there exists a relatively dense set of numbers $\{\tau\}$ such that

$$|\varphi(x+\tau) - \varphi(x)| < \varepsilon \tag{7.21}$$

for any τ from this set and for all $x \in R$.

THEOREM 5.21.   *A necessary and sufficient condition for the motion $f_u(\varphi, t)$ in the Bebutov dynamical system to be almost periodic is that the function $\varphi(x)$ be Bohr almost periodic.*

Proof. Suppose the motion $f_u(\varphi, t)$ is almost periodic. Then for any $\varepsilon > 0$ there exists a relatively dense set of points $\{\tau\}$ such that

$$\rho(f_u(\varphi, t+\tau), f_u(\varphi, t)) < \varepsilon$$

for all $x \in R$. Furthermore, according to Lemma 1.20,

$$\max_{|x| \leq 1/\varepsilon} |\varphi(x+t+\tau) - \varphi(x+t)| \leqq \varepsilon \qquad (8.21)$$

for all $t \in R$. Setting $x = 0$, we obtain from (8.21) that

$$|\varphi(t+\tau) - \varphi(t)| \leqq \varepsilon$$

for all $t \in R$ and any τ from the relatively dense set $\{\tau\}$; but this means that the function $\varphi(x)$ is Bohr almost periodic.

Now assume that the function $\varphi(x)$ is Bohr almost periodic. Inasmuch as under this assumption inequality (7.21) holds for any τ from the set $\{\tau\}$ and for all $x \in R$, we also have that

$$|\varphi(x+t+\tau) - \varphi(x+t)| < \varepsilon$$

for any $\tau \in \{\tau\}$ and all $x, t \in R$. Then condition (8.21) is satisfied and, by Lemma 1.20,

$$\rho(f_u(\varphi, t+\tau), f_u(\varphi, t)) \leqq \varepsilon;$$

hence the motion $f_u(\varphi, t)$ is almost periodic.

§ 22   Lyapunov stability

Definition 1.22.   A point p and motion $f(p, t)$ are *positively Lyapunov stable* if for any $\varepsilon > 0$ there exists a $\delta > 0$ such that

$$\rho(f(p, t), f(q, t)) < \varepsilon \qquad (1.22)$$

provided $\rho(p, q) < \delta$ and $t \in R^+$.

Clearly, the condition that the motion $f(p, t)$ be positively Lyapunov stable is nothing other than the condition of integral continuity (see Theorem 4.2) with $T = +\infty$ (i.e., on an infinite interval of time).

Definition 2.22.   A point $p \in X$ and motion $f(p, t)$ are *positively Lyapunov stable relative to the set* $B \subseteq X$ if $p \in \bar{B}$ and for any $\varepsilon > 0$ there exists a $\delta > 0$ such that inequality (1.22) is satisfied provided $\rho(p, q) < \delta$, $q \in B$, $t \in R^+$.

Obviously, for $B = X$ Definition 2.22 reduces to Definition 1.22. It is

95

obvious that if the point p is positively Lyapunov stable relative to the set B then $f(p, R)$ may have points which are not positively Lyapunov stable relative to the set B if only because they may not belong to the set $\bar{B}$. However, the following theorem does hold.

THEOREM 1.22. *The set of points which are positively Lyapunov stable relative to an invariant set B is invariant.*

Proof. Suppose that the set B is invariant and that the point $p \in X$ is positively Lagrange stable relative to B. Then $p \in \bar{B}$ and for every $\varepsilon > 0$ there exists a $\delta > 0$ such that the inequality (1.22) is satisfied for all $q \in B$ such that $\rho(p, q) < \delta$ and all $t > 0$. We may assume that $\delta < \varepsilon$.

Consider an arbitrary point $p_1 = f(p, t_1)$ of the trajectory $f(p, R)$. We shall show that the point p_1 is positively Lagrange stable relative to B. According to Theorem 5.4 the point $p_1 \in \bar{B}$. For $\delta > 0$, the point p_1 and the numbers $|t_1|$, from the condition of integral continuity there exists an $\eta > 0$ such that for $\rho(r, p_1) < \eta$ and all $|t| \leq |t_1|$,

$$\rho(f(r, t), f(p_1, t)) < \delta. \tag{2.22}$$

We shall show that this η is a suitable choice for $\delta > 0$ in the condition that the point p_1 be positively Lagrange stable relative to B. To this end, take an arbitrary point $r \in B$ for which $\rho(r, p_1) < \eta$. Then inequality (2.22) is satisfied for all $|t| \leq |t_1|$ and in particular for $t = -t_1$. Thus

$$\rho(f(r, -t_1), p) < \delta$$

and furthermore

$$f(r, -t_1) \equiv q \in B$$

by virtue of the invariance of B. Then by definition, for all $t > 0$, δ satisfies inequality (1.22), i.e., (inasmuch as $p = f(p_1, -t_1)$),

$$\rho(f(r, t-t_1), f(p_1, t-t_1)) < \varepsilon$$

for all $t > 0$, i.e.,

$$\rho(f(r, t), f(p_1, t)) < \varepsilon \tag{3.22}$$

for all $t > -t_1$. Taking (2.22) and the fact that $\delta < \varepsilon$ into consideration, we conclude that inequality (3.22) holds for all $|t| \leq |t_1|$ and thus it is satisfied for all $t > 0$. Therefore the point p_1 is positively Lagrange stable relative to B, which is what was required to be proved.

Definition 3.22.   A set $A \subseteq X$ is said to be *positively Lyapunov stable relative to the set* $B \subseteq X$ if every point of the set A is positively Lyapunov stable relative to the set B (and furthermore $A \subseteq \bar{B}$).

Definition 4.22.   A set $A \subseteq X$ is said to be *uniformly positively Lyapunov stable relative to the set* B if $A \subseteq \bar{B}$ and for any $\varepsilon > 0$ there exists a $\delta(\varepsilon) > 0$ such that inequality (1.22) holds provided $\rho(p, q) < \delta$, $p \in A$, $q \in B$ and $t \in R^+$.

THEOREM 2.22.   *If the set A is (uniformly) positively Lyapunov stable relative to the set B, then A is (uniformly) positively Lyapunov stable relative to $\bar{B}$.*

Proof. Let $p \in A$, $\varepsilon > 0$. Then $p \in \bar{B}$ and there exists a $(\delta = \delta(\varepsilon/2) > 0)$ $\delta = \delta(\varepsilon/2, p) > 0$ such that

$$\rho(f(p, t), f(r, t)) < \frac{\varepsilon}{2} \tag{4.22}$$

provided

$$p \in A, r \in S(p, \delta) \cap B, t \in R^+.$$

We shall show that the number $\delta/2$ is a suitable choice for δ in the condition that the (set A) point p be positively Lyapunov stable relative to the set $\bar{B}$. Suppose that

$$p \in A, q \in S\left(p, \frac{\delta}{2}\right) \cap \bar{B}, t \in R^+$$

By the continuity condition 2.1, for the point q and the numbers t and $\varepsilon/2$ there exists an $\eta > 0$ such that for

$$\rho(r, q) < \eta, \tag{5.22}$$

$$\rho(f(r, t), f(q, t)) < \frac{\varepsilon}{2}. \tag{6.22}$$

We may assume that $\eta < \delta/2$. Fix the point $r \in B$ so that inequality (5.22) is satisfied. Furthermore $\rho(r, p) < \delta$ and inequalities (4.22) and (6.22) are satisfied, from which (1.22) follows, which is what was required to be proved.

THEOREM 3.22.   *If the set A is uniformly positively Lyapunov stable relative to the set B, then $\bar{A}$ is uniformly positively Lyapunov stable relative to the set $\bar{B}$.*

Proof. According to Theorem 2.22 the set A is uniformly positively Lyapunov stable relative to the set $\bar{B}$. Let $\varepsilon > 0$. Then $A \subseteq \bar{B}$ (and $\bar{A} \subseteq \bar{B}$) and there exists a $\delta = \delta(\varepsilon/2) > 0$ such that inequality (6.22) is satisfied provided

$$t \in R^+, r \in A, q \in \bar{B}, \rho(r, q) < \delta.$$

We shall prove that the number $\delta/2$ is a suitable choice for δ in the condition for the set $\bar{A}$ to be uniformly positively Lyapunov stable relative to the set $\bar{B}$. Assume that $p \in \bar{A}$, $q \in \bar{B}$ and that

$$\rho(p, q) < \frac{\delta}{2}. \tag{7.22}$$

For the point p and the numbers $\varepsilon/2$ and t there exists by the continuity condition 2.1 an $\eta > 0$ such that for any point $r \in S(p, \eta)$ inequality (4.22) is satisfied.

Since $p \in \bar{A}$ there exists a point $r \in A$ such that

$$\rho(p, r) < \min\left(\frac{\delta}{2}, \eta\right). \tag{8.22}$$

Then, by virtue of (7.22) and (8.22), $\rho(r, q) < \delta$ and inequalities (6.22) and (4.22) must be satisfied for the chosen point $r \in A$. These inequalities imply (1.22), which is what was required to be proved.

Simple examples show that if A is positively Lyapunov stable relative to $\bar{B}$ then $\bar{A}$ may not be positively Lyapunov stable relative to $\bar{B}$.

THEOREM 4.22.   *If the compact set A is positively Lyapunov stable relative to the set B, then A is uniformly positively Lyapunov stable relative to $\bar{B}$.*

Proof. According to Theorem 2.22, the set A is positively Lyapunov stable relative to $\bar{B}$. Let $\varepsilon > 0$. By assumption $A \subseteq \bar{B} \subseteq X$ and for each point $p \in A$ there exists a $\delta(p, \varepsilon/2)$ such that inequality (4.22) is satisfied provided

$$\rho(p, r) < \delta\left(p, \frac{\varepsilon}{2}\right), r \in \bar{B}, t \in R^+.$$

Consider the collection of neighborhoods

$$\left\{ S\left(p, \tfrac{1}{2}\delta\left(p, \frac{\varepsilon}{2}\right)\right) \right\}$$

which form a cover of A. Since A is compact, this collection contains a finite subcover:

$$S_i = S\left(p_i, \tfrac{1}{2}\delta\left(p_i, \frac{\varepsilon}{2}\right)\right) \quad (i = 1, 2, \ldots, n)$$

[see P. S. Aleksandrov [2], page 318, Theorem 6 (Borel-Lebesgue)].
Set

$$\delta_i = \delta\left(p_i, \frac{\varepsilon}{2}\right), \ \delta = \min_i \left\{\frac{\delta_i}{2}\right\}.$$

We shall show that this δ is a suitable choice for the δ in the definition of the uniform positive Lyapunov stability of the set A relative to the set $\bar{B}$. To this end, take an arbitrary $t \in R^+$ and two points $p \in A$ and $q \in \bar{B}$ such that

$$\rho(p, q) < \delta. \tag{9.22}$$

Since $p \in A$, there exists a neighborhood S_k such that $p \in S_k$, i.e.,

$$\rho(p_k, p) < \tfrac{1}{2}\delta_k. \tag{10.22}$$

It follows from (9.22) and (10.22), by virtue of the fact that $\delta \leq \tfrac{1}{2}\delta_k$, that

$$\rho(p_k, q) < \delta_k,$$

and then, according to the definition of δ_k,

$$\rho(f(p_k, t), f(q, t)) < \frac{\varepsilon}{2}. \tag{11.22}$$

Since $p \in A \subseteq \bar{B}$, it follows from (10.22), according to the choice of δ_k, that also

$$\rho(f(p_k, t), f(p, t)) < \frac{\varepsilon}{2}. \tag{12.22}$$

Now (1.22) follows from (11.22) and (12.22), which proves that the set A is uniformly positively Lyapunov stable relative to the set $\bar{B}$, which is what was required to be proved.

Analogous definitions and theorems can be stated and proved for negatively Lyapunov stable motions and also for motions which are both positively and negatively Lyapunov stable. In the first case the corresponding inequalities must be satisfied for all $t \in R^-$ and in the second case for all real t.

THEOREM 5.22.   *If the set Σ_q is Lyapunov stable relative to $f(p, R)$, then Σ_p is a minimal set.*

Proof. Let $q \in \Sigma_p$. Then $\Sigma_q \subseteq \Sigma_p$. The point q is Lyapunov stable relative to $f(p, R)$ and therefore for any $\varepsilon > 0$ there exists a $\delta > 0$ such that

$$\rho(f(r, t), f(q, t)) < \varepsilon \tag{13.22}$$

provided $t \in R$, $r \in f(p, R)$ and $\delta(r, q) < \delta$. Fix the point $r = f(p, t_0) \in S(q, \delta)$ and the moment $t = -t_0$. Then from (13.22) we obtain that

$$\rho(p, f(q, -t_0)) < \varepsilon.$$

Therefore $p \in \Sigma_q$ so that then $\Sigma_p \subseteq \Sigma_q$. It follows that $\Sigma_q = \Sigma_p$ and by Theorem 2.16 we conclude that Σ_p is a minimal set.

Theorems 3.22 and 5.22 imply the following corollary.

COROLLARY 1.22. *If the set $f(p, R)$ is uniformly Lyapunov stable relative to $f(p, R)$, then Σ_p is a minimal set.*

THEOREM 6.22 (DEYSACH AND SELL [1]). *If the motion $f(p, t)$ is positively Lagrange stable and positively asymptotically stable, then $f(p, R^+)$ is positively Lyapunov stable relative to $f(p, R^+)$.*

Proof. Let $q \in f(p, R^+)$, i.e., let $q = f(p, s)$ where $s \in R^+$. We shall prove that the point q is positively Lyapunov stable relative to $f(p, R^+)$. Suppose that q is not positively Lyapunov stable relative to $f(p, R^+)$. Then there exist an $\varepsilon_0 > 0$ and sequences

$$\{p_n\} \subseteq f(p, R^+), \{t_n\} \subseteq R^+$$

such that $\{p_n\} \to q$ whereas

$$\rho(f(q, t_n), f(p_n, t_n)) \geqq \varepsilon. \tag{14.22}$$

Since $p_n \in f(p, R^+)$, then $p_n = f(p, s_n)$ where $s_n \geqq 0$. Inasmuch as $f(p, t)$ is positively asymptotically stable, the point $q \notin \Omega_p$. Therefore the sequence $\{s_n\}$ is bounded. We may therefore assume that $\{s_n\} \to s_0$. Then

$$\{p_n\} = \{f(p, s_n)\} = \{f(q, s_n - s)\} \to f(q, s_0 - s).$$

But $\{p_n\} \to q$ and therefore

$$f(q, s_0 - s) = q.$$

Since the motion $f(p, t)$ is positively asymptotically stable, it cannot be periodic and therefore $s_0 = s$. By virtue of the positive Lagrange stability, the sequence $\{f(p, t_n)\}$ can be assumed convergent. Let

$$\{f(p, t_n)\} \to r.$$

Then

$$\{f(p_n, t_n)\} = \{f(p, t_n + s_n)\} = \{f(f(p, t_n), s_n)\} \to f(r, s)$$

and

$$\{f(q, t_n)\} = \{f(p, t_n + s)\} \to f(r, s),$$

which contradicts inequality (14.22).

§ 23   Lyapunov stable motions on the real line

Let $f(x, t)$ be a dynamical system defined on the Euclidean line $R = R^1$. We denote by N the set of points which are positively Lyapunov stable and by $M = R^1 \setminus N$ the set of points which are not positively Lyapunov stable. On the basis of Theorem 1.22 both of these sets are invariant.

THEOREM 1.23.   *If x_0 is not a rest point and the motion $f(x_0, t)$ is positively (negatively) Lagrange stable, then it is positively (negatively) Lyapunov stable.*
     Proof. Let

$$\Omega_{x_0} = b(|b| < +\infty), A_{x_0} = a(|a| \leqq +\infty, a < b).$$

Take any point $x_1 \in (a, b)$, $x_1 < x_0$. By Theorem 6.22, the point x_0 is positively Lyapunov stable relative to $f(x_1, R^+)$; but now it is clear that x_0 is positively Lyapunov stable, i.e., $x_0 \in N$, which is what was required to be proved.

    In the case of motions on the real line which are not Lagrange stable nothing can be said *a priori* concerning whether or not they are Lyapunov stable. Thus, in Example 5.6, the motions are Lyapunov stable, whereas in Example 6.6 it follows from formula (2.6) that if $y_0 > 0$, $y_0 \neq x_0$, $t > 0$ and $y = (1 + y_0)e^t - 1$, then, as $t \to +\infty$,

$$x - y = (x_0 - y_0)e^t \to \infty,$$

and the motion is not Lyapunov stable despite the fact that the dynamical systems of Examples 5.6 and 6.6 are isomorphic by Lemma 3.6.
    Definition 1.23.   We shall say that a set $Q \subseteq R^1$ is a *set of one-sided isolated points* if for each point $x \in Q$ there exists a number $\delta > 0$ such that at least one of the intervals $(x - \delta, x)$ and $(x, x + \delta)$ does not contain points of Q.
    The following theorem completely characterizes the set M.

THEOREM 2.23 (I. VRKOČ [2]).   *A necessary and sufficient condition for a set of points $M \subseteq R^1$ to be for some dynamical system the set of points which are not positively Lyapunov stable is that it consist of not more than a countable set Q of one-sided isolated points and not more than two infinite closed intervals.*

Proof. Necessity. Suppose given a dynamical system on R^1 for which M is the set of points which are not positively Lyapunov stable. If there are no rest points then, since M is invariant, the set $M = R^1$ and consists of the single infinite closed interval $(-\infty, +\infty)$. But if the set of rest points $K \neq \phi$, then we set

$$a = \inf K, \quad b = \sup K.$$

Let $[b, +\infty) \neq \phi$. In this situation, $b \in K$ and $(b, +\infty)$ consists of one trajectory. In this connection, since M is invariant, either

$$(b, +\infty) \subseteq M$$

or

$$(b, +\infty) \cap M = \phi. \tag{1.23}$$

In the first case, taking Theorem 1.23 into consideration, we conclude that the motions on $(b, +\infty)$ must occur from left to right. In this connection, also the rest point $b \in M$, i.e.,

$$[b, +\infty) \subseteq M.$$

An analogous line of reasoning can be followed in the case when $(-\infty, a] \neq \phi$.

Now consider (a, b) and set

$$(a, b) \cap M \equiv Q.$$

On the basis of Theorem 1.23 the set $Q \subseteq K$. Let $x \in Q$. It is clear that if for every $\delta > 0$ both of the intervals $(x - \delta, x)$ and $(x, x + \delta)$ contained points from Q (and they are rest points), then the point x would be Lyapunov stable. Consequently, there exists a $\delta > 0$ for which at least one of these two intervals does not contain points from Q. Thus, to each point $x \in Q$ there is set into correspondence at least one of the intervals $(x - \delta, x)$ and $(x, x + \delta)$. Since these intervals can be chosen to be disjoint, then Q is at most denumerable. Add the point b to the set Q if $b < +\infty$ and (1.2) is satisfied, and add the point a if $a > -\infty$ and

$$(-\infty, a) \cap M = \phi.$$

Sufficiency. Let M be a given set consisting of at most a denumerable collection Q of one-sided isolated points and not more than two infinite closed intervals. We can assume that these intervals and the set Q are pairwise disjoint. We define on the infinite closed intervals belonging to M, if there are any, a positively departing motion which is not positively Lyapunov stable, and we declare the endpoints of these intervals and the points of the set $\bar{Q}$, if it is nonempty, to be rest points. Then the dynamical system is already defined on the set $\bar{M}$.

Let (ξ, η) be an interval adjacent to $\bar{M}$. Then ξ and η are rest points. Note that by virtue of the properties of the set Q each of its points is an endpoint of an interval which is adjacent to $\bar{M}$. Here, the following four cases are possible:

1) $\xi \in M$ and $\eta \in M$;   2) $\xi \in M$ and $\eta \notin M$;

3) $\xi \notin M$ and $\eta \in M$;   4) $\xi \notin M$ and $\eta \notin M$.

In case 1) we declare the point $(\xi + \eta)/2$ to be a rest point; in $(\xi, (\xi + \eta)/2)$ we define the motion to be in the direction from left to right and in $((\xi + \eta)/2, \eta)$ in the opposite direction.

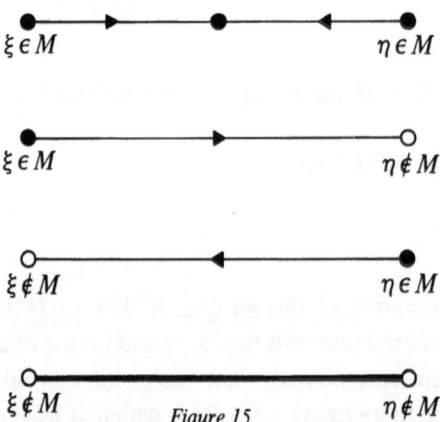

Figure 15

In case 2), we define the motion in (ξ, η) to be from the left to the right; in case 3) from the right to the left. In case 4) we declare all points of the interval (ξ, η) to be rest points (see Fig. 15). Thus it is clear that a dynamical system with the given set M of points which are not positively Lyapunov stable has been constructed in R^1. Furthermore, the set of rest points is obviously closed as should be the case.

§ 24   Interrelationship between periodicity and Lyapunov stability

Definition 1.24.   A dynamical system is *almost periodic* if for any $\varepsilon > 0$ there exists a relatively dense set of ε-displacements of the space X.

Definition 2.24.   A dynamical system is *Lyapunov stable* if all points of the space X are Lyapunov stable.

THEOREM 1.24.   *An almost periodic dynamical system is Lyapunov stable.*

Proof. Take an arbitrary $\varepsilon > 0$ and $p \in X$. By virtue of the almost periodicity, for $\varepsilon/3$ there exists a $T > 0$ such that on any segment of time of length T there exists an $(\varepsilon/3)$-displacement of the space X, i.e., an $(\varepsilon/3)$-displacement of all points of X. For the point p and the numbers $\varepsilon/3$ and T there exists a corresponding $\delta > 0$ from the integral continuity condition such that

$$\rho(f(p, t), f(q, t)) < \frac{\varepsilon}{3} \tag{1.24}$$

provided $\rho(p, q) < \delta$ and $0 \leqq t \leqq T$. Now take an arbitrary point $q \in X$ such that $\rho(p, q) < \delta$. We shall prove that condition (1.22) is satisfied for all t. In fact, let t be an arbitrary number. Then in virtue of (1.24)

$$\rho(f(p, t+\tau), f(q, t+\tau)) < \frac{\varepsilon}{3} \tag{2.24}$$

if $0 \leqq t+\tau \leqq T$, i.e., $-t \leqq \tau \leqq -t+T$. On $[-t, -t+T]$ we can take, by the definition of T, as τ some $(\varepsilon/3)$-displacement for all points of the space X. Then, besides inequality (2.24), also the inequalities

$$\rho(f(p, t+\tau), f(p, t)) < \frac{\varepsilon}{3} \tag{3.24}$$

and

$$\rho(f(q, t+\tau), f(q, t)) < \frac{\varepsilon}{3} \tag{4.24}$$

will hold. Now (1.22) follows from (3.24), (2.24) and (4.24) and therefore the point p and the motion $f(p, t)$ are Lyapunov stable. This completes the proof of the theorem.

According to Remark 1.21, a dynamical system which is induced on the closure of a trajectory of an almost periodic motion is almost periodic. Taking this into account, the next corollary follows from Theorem 1.24.

COROLLARY 1.24.   *The closure Σ_p of the trajectory of an almost periodic motion $f(p, t)$ is Lyapunov stable relative to Σ_p.*

This corollary and Theorem 4.22 imply the next corollary.

COROLLARY 2.24.   *If $f(p, t)$ is a Lagrange stable almost periodic motion, the set Σ_p is uniformly Lyapunov stable relative to Σ_p.*

This corollary and Theorem 3.17 imply the next corollary.

COROLLARY 3.24.   *In a complete space, the minimal set Σ of almost periodic motions is uniformly Lyapunov stable relative to Σ.*

These corollaries contain conditions under which the almost periodicity of $f(p, t)$ implies simple or uniform Lyapunov stability. One must note however that the uniform Lyapunov stability relative to the set Σ_p of the set Σ_p does not imply the almost periodicity of $f(p, t)$. Thus, in the case of a uniform motion on the real line (Example 5.6), Σ_p is uniformly Lyapunov stable relative to the set Σ_p but $f(p, t)$ is not almost periodic (since it is not Poisson stable).

A. A. Markov [2] established certain conditions under which even one-sided Lyapunov stability implies almost periodicity. These conditions are contained in the following two theorems.

THEOREM 2.24.   *If the motion $f(p, t)$ is almost recurrent (negatively Poisson stable) and positively Lyapunov stable relative to $f(p, R)$, then it is almost periodic (uniformly Poisson stable).*

Proof. Let $\varepsilon > 0$. From the condition of the positive Lyapunov stability relative to $f(p, R)$ of the point p, we find the corresponding $\delta = \delta(\varepsilon/2) > 0$.

The almost recurrence (negative Poisson stability) implies the existence of an (unbounded) relatively dense set of numbers $\{\tau\}$ which satisfy the condition

$$\rho(p, f(p, \tau)) < \frac{\delta}{2}. \qquad (5.24)$$

Let $t \in R$ and $\tau \in \{\tau\}$. We shall prove that τ is an ε-displacement of the point $f(p, t)$. For the given τ, there exists by the continuity condition 2.1 an $\eta > 0$ such that the inequality $\rho(p, q) < \eta$ implies

$$\rho(f(p, \tau), f(q, \tau)) < \frac{\delta}{2}.$$

Since the motion $f(p, t)$ is negatively Poisson stable, there exists a $t_1 < t$ such that

$$\rho(p, f(p, t_1)) < \min (\eta, \delta), \tag{6.24}$$

and then by the definition of η we have that

$$\rho(f(p, \tau), f(p, t_1 + \tau)) < \frac{\delta}{2}. \tag{7.24}$$

It follows from (5.24) and (7.24) that

$$\rho(p, f(p, t_1 + \tau)) < \delta. \tag{8.24}$$

Since $t - t_1 > 0$, then, corresponding to the choice of δ, the inequalities

$$\rho(f(p, t - t_1), f(p, t)) < \frac{\varepsilon}{2} \tag{9.24}$$

and

$$\rho(f(p, t - t_1), f(p, t + \tau)) < \frac{\varepsilon}{2} \tag{10.24}$$

follow from (6.24) and (8.24). From (9.24) and (10.24) we obtain that

$$\rho(f(p, t), f(p, t + \tau)) < \varepsilon,$$

which is what was required to be proved (see Remark 2.21).

COROLLARY 4.24. *If an almost recurrent motion $f(p, t)$ is Lyapunov stable relative to $f(p, R)$, then it is almost periodic.*

THEOREM 3.24. *If $f(p, R)$ is uniformly positively Lyapunov stable relative to $f(p, t)$ and the motion $f(p, t)$ is negatively Lagrange stable, then $f(p, t)$ is almost periodic.*

Proof. By virtue of Theorem 2.24, it suffices to prove that $f(p, t)$ is almost recurrent. Assume the contrary. Since $f(p, t)$ is negatively Lagrange stable, the set A_p is compact and it contains a compact minimal set Σ:

$$\Sigma \subseteq A_p \subseteq \Sigma_p.$$

Furthermore, we may assume that Σ is a proper subset (because otherwise $\Sigma = \Sigma_p = A_p$ and by Theorem 2.18 all motions in Σ_p are recurrent). It is clear that the point p does not occur in Σ and therefore

$$\rho(p, \Sigma) \equiv \alpha > 0. \tag{11.24}$$

By Theorem 3.22 the set Σ_p is uniformly positively Lyapunov stable relative to Σ_p. From this condition there exists a corresponding $\delta = \delta(\alpha/2) > 0$.

Let $q \in \Sigma$. Since $\Sigma \subseteq A_p$, then there exists a $t_0 > 0$ for which the point

$$p^* \equiv f(p, -t_0) \in S(q, \delta).$$

Then, corresponding to the choice of δ,

$$\rho(f(p^*, t_0), f(q, t_0)) = \rho(p, f(q, t_0)) < \frac{\alpha}{2},$$

and since $f(q, t_0) \in \Sigma$ we have that

$$\rho(p, \Sigma) < \frac{\alpha}{2},$$

which contradicts (11.24). This completes the proof of the theorem.

Theorems 3.24 and 4.22 imply the following corollary.

COROLLARY 5.24. *If the motion $f(p, t)$ is Lagrange stable and Σ_p is Lyapunov stable relative to Σ_p, then the motion $f(p, t)$ is almost periodic.*

Corollaries 2.24 and 5.24 can be combined in the form of the following theorem.

THEOREM 4.24. *A necessary and sufficient condition for a Lagrange stable motion $f(p, t)$ to be almost periodic is that Σ_p be (uniformly) Lyapunov stable relative to Σ_p.*

It is convenient to depict schematically many of the established properties of a motion of a dynamical system as follows:

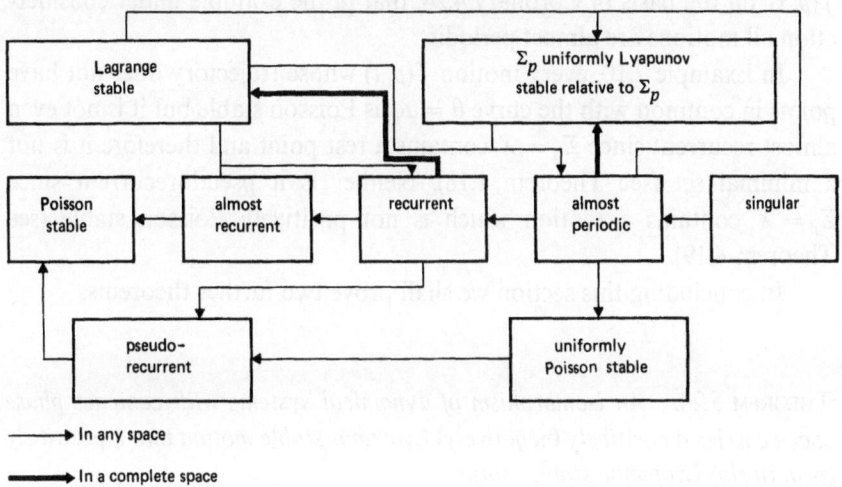

By virtue of Theorem 1.9, for a dynamical system defined in the plane, all almost periodic, all recurrent motions as well as all Poisson stable motions are exhausted by singular motions.

In Example 1.10, all motions are recurrent for α irrational (see the first theorem of Birkhoff) since in this case $\Sigma_p = X$ for any point $p \in X$ and consequently the entire torus is a compact minimal set. We shall prove that in this example the dynamical system is uniformly Lyapunov stable (i.e., X is uniformly Lyapunov stable relative to X).

In fact, if we consider two arbitrary points p_1 and p_2 with coordinates (φ_1, θ_1) and (φ_2, θ_2) on the torus, then the points $f(p_1, t)$ and $f(p_2, t)$ will have the coordinates

$$(\varphi_1 + t, \theta_1 + \alpha t), (\varphi_2 + t, \theta_2 + \alpha t),$$

respectively. Since in the definition of the distance between two points on the torus there occur the differences between corresponding coordinates, then it is clear that

$$\rho(f(p_1, t), f(p_2, t)) = \rho(p_1, p_2).$$

From this it is clear that we can take $\delta = \varepsilon$ in the condition for the uniform Lyapunov stability of X relative to X.

It follows from the recurrence and Lyapunov stability of every motion $f(p, t)$, on the basis of Corollary 4.24, that in the example under consideration all motions are almost periodic.

In Example 2.10, every motion $f(p, t)$ whose trajectory does not have points in common with the curve $\theta = \alpha\varphi$ is Poisson stable but it is not even almost recurrent since $\Sigma_p = X$ contains a rest point and therefore it is not a minimal set (see Theorem 1.18). Neither is it pseudorecurrent since $\Sigma_p = X$ contains a motion which is not positively Poisson stable (see Theorem 4.19).

In concluding this section we shall prove two further theorems.

THEOREM 5.24.   *An isomorphism of dynamical systems with compact phase space carries a positively (negatively) Lyapunov stable motion into a positively (negatively) Lyapunov stable motion.*

Proof. Let $\xi: X_1 \to X_2$ be a homeomorphism which realizes the isomorphism of the dynamical systems $g(p, t)$ and $h(x, t)$. In view of the compactness of the spaces X_1 and X_2 the mappings ξ and ξ^{-1} must be uniformly continuous. Let $\varepsilon > 0$. There exists an $\eta > 0$ such that the inequality $\rho(\xi(q), \xi(p)) < \varepsilon$ is satisfied provided $\rho(p, q) < \eta$ $(p, q \in X_1)$.

Now assume that the motion $g(p, t)$ is positively Lyapunov stable. Then corresponding to the number $\eta > 0$ there exists a $\delta > 0$ such that

$$\rho(g(p, t), g(q, t)) < \eta \qquad\qquad (12.24)$$

provided $\rho(p, q) < \delta$ $(p, q \in X_1)$ and $t \in R^+$.

Finally, corresponding to the number $\delta > 0$ there exists a $\kappa > 0$ such that

$$\rho(\xi^{-1}(x), \xi^{-1}(y)) < \delta \qquad\qquad (13.24)$$

provided

$$\rho(x, y) < \kappa \ (x, y \in X_2).$$

Let $x = \xi(p)$, $\rho(x, y) < \kappa(x, y \in X_2)$ and $t > 0$. Then inequality (13.24) is satisfied. Setting $\xi^{-1}(y) = q$, we obtain that $\rho(p, q) < \delta$ and consequently inequality (12.24) holds. Furthermore, according to the choice of η,

$$\rho(\xi(g(p, t)), \xi(g\ (q, t))) < \varepsilon,$$

and, according to (1.6),

$$\rho(h(\xi(p), t), h(\xi(q), t)) < \varepsilon,$$

i.e.,

$$\rho(h(x, t), h(y, t)) < \varepsilon$$

and the motion $h(x, t)$ is positively Lyapunov stable, which is what was to be proved.

We note that the theorem analogous to the one just proved in the case of a homomorphism is not valid. This fact is easily discovered by the following example. Let $g(p, t)$ be a dynamical system defined on a segment $X_1 \equiv [a, b] \subset R^1$ so that a and b are rest points and the interval (a, b) is the trajectory of some motion. For the space X_2 take a circle of arc length $b - a$. We define the system in X_2 by the homomorphism which wraps the segment $[a, b]$ onto the circle X_2. Let x_0 be the point in X_2 which is the point into which the points a and b coalesce in this process. The system $h(x, t)$ will be a homomorphic image of the system $g(p, t)$. Each of the points a and b is either positively or negatively Lyapunov stable whereas the point x_0 is neither positively nor negatively Lyapunov stable.

THEOREM 6.24.   *A homomorphism of dynamical systems with a complete (compact) phase space X_1 carries an almost periodic (uniformly Poisson stable) motion into an almost periodic (uniformly Poisson stable) motion.*

Proof. Let $\xi: X_1 \rightarrow X_2$ be a continuous mapping which realizes the homomorphism of the dynamical system $g(p, t)$ into $h(x, t)$. Taking Theorem 3.17 into consideration, we can assume the space X_1 to be compact. Then the mapping ξ is uniformly continuous and for any $\varepsilon > 0$ there exists an $\eta > 0$ such that the inequality $\rho(\xi(p), \xi(q)) < \varepsilon$ is satisfied provided $\rho(p, q) < \eta$ $(p, q \in X_1)$.

Now assume that the motion $g(p, t)$ is almost periodic (uniformly

Poisson stable) and let $x = \xi(p)$. Then for the number $\eta > 0$ there exists a relatively dense (unbounded) set of numbers $\{\tau\}$ such that

$$\rho(g(p, t+\tau), g(p, t)) < \eta$$

for all $t \in R$, $\tau \in \{\tau\}$. It follows from this, according to the choice of η, that

$$\rho(\xi(g(p, t+\tau)), \xi(g(p, t))) < \varepsilon,$$

and, according to (1.6),

i.e.,
$$\rho(h(\xi(p), t+\tau), h(\xi(p), t)) < \varepsilon,$$
$$\rho(h(x, t+\tau), h(x, t)) < \varepsilon$$

for all $t \in R$, $\tau \in \{\tau\}$, and the motion $h(x, t)$ is almost periodic (uniformly Poisson stable). This completes the proof of the theorem.

§ 25   Motions in dynamical limit sets

The basic results of this section are due to the author [1, 2].

We consider the question of the influence of various properties of a motion on the motions in their dynamical limit sets.

Definition 1.25. A semitrajectory $f(p, R^+)[f(p, R^-)]$ *uniformly approximates the set Q* if for any $\varepsilon > 0$ there exists a $T(\varepsilon) > 0$ such that any arc of the semitrajectory $f(p, R^+)[f(p, R^-)]$ of time length T approximates the set Q to within ε, i.e.,

$$Q \subseteq S(f(p, [t_0, t_0 + T]), \varepsilon)$$

for any $t_0 \geq 0 \, (t_0 \leq -T)$.

Definition 2.25. A point q is a $\psi(\beta)$-*limit point of the motion $f(p, t)$* if the semitrajectory $f(p, R^+)[f(p, R^-)]$ uniformly approximates the point q. The set of all $\psi(\beta)$-limit points of the motion $f(p, t)$ is called the $\psi(\beta)$-*limit set* of this motion and is denoted by $\Psi_p (B_p)$.

In the sequel we shall deal only with the sets Ω_p and Ψ_p. All the results can be easily carried over to the sets A_p and B_p.

Definition 3.25. A sequence of nonnegative numbers $\{t_n\}$ is *relatively dense on* R^+ if there exists an $L > 0$ such that for any $l \in R^+$ at least one point of this sequence occurs on the segment $[l, l+L]$.

It is clear that the following lemma holds.

LEMMA 1.25. *The point* q *is a* ψ-*limit point of the motion* $f(p, t)$ *if and only if for any* $\varepsilon > 0$ *there exists a relatively dense sequence* $\{t_n\}$ *on* R^+ *such that*

$$\bigcup_{n=1}^{\infty} f(p, t_n) \subseteq S(q, \varepsilon). \tag{1.25}$$

It follows from this that
$$\Psi_p \subseteq \Omega_p \subseteq \Sigma_p^+. \tag{2.25}$$

Clearly, if the motion $f(p, t)$ is singular, then

$$\Psi_p = \Omega_p = f(p, R).$$

THEOREM 1.25. *The set* Ψ_p *is invariant.*

Proof. Let $q_0 \in \Psi_p$. Take an arbitrary point

$$q = f(q_0, \tau) \in f(q_0, R)$$

and a number $\varepsilon > 0$. By the continuity condition for a dynamical system there exists a $\delta > 0$ such that for $\rho(r, q_0) < \delta$, we have

$$\rho(f(r, \tau), f(q_0, \tau)) < \varepsilon.$$

From the condition $q_0 \in \Psi_p$ we can find a relatively dense sequence $\{t_n\}$ on R^+ such that

$$\bigcup_{n=1}^{\infty} f(p, t_n) \subseteq S(q_0, \delta).$$

Then

$$\bigcup_{n=1}^{\infty} f(p, t_n + \tau) \subseteq S(q, \varepsilon).$$

It is easy to see that $\{t_n+\tau\} \cap R^+$ is a relatively dense sequence on R^+ and therefore, according to Lemma 1.25, $q \in \Psi_p$. This completes the proof of the theorem.

THEOREM 2.25. *The set Ψ_p is closed.*
  Proof. Suppose $q_0 \in \bar{\Psi}_p$ and that $\varepsilon_0 > 0$. There exist a point $q \in \Psi_p$ and a number $\varepsilon > 0$ such that

$$S(q, \varepsilon) \subseteq S(q_0, \varepsilon_0).$$

For this number ε there exists a relatively dense sequence $\{t_n\}$ on R^+ for which condition (1.25) is satisfied. Then

$$\bigcup_{n=1}^{\infty} f(p, t_n) \subseteq S(q_0, \varepsilon_0)$$

and the point $q_0 \in \Psi_p$.

COROLLARY 1.25. *If K is a nonempty subset of Ψ_p, then $\Sigma_K \subseteq \Psi_p$.*

THEOREM 3.25. *The set Ψ_p is empty in each of the following cases:*
  1. Ω_p *contains more than one minimal set.*
  2. *The space X is locally compact and the motion $f(p, t)$ is not positively Lagrange stable.*
  We preface the proof of this theorem by two lemmas.

LEMMA 2.25. *If Ω_p contains a minimal set M, then either $\Psi_p = M$ or $\Psi_p = \phi$.*
  Proof. Suppose there exists a minimal set $M \subseteq \Omega_p$. Then, according to Lemma 1.18, the motion $f(p, t)$ spends for $t > 0$ arbitrarily large intervals of time in any neighborhood of the set M and therefore no other point except the points in the set M can belong to Ψ_p, i.e., $\Psi_p \subseteq M$.
  Assume that $\Psi_p \neq \phi$. Then, being a closed invariant subset of the minimal set M, Ψ_p must coincide with M. This completes the proof of the lemma.

LEMMA 3.25. *If a point of local compactness of the space X belongs to the ψ-limit set of a motion, then that motion is positively Lagrange stable.*
  Proof. Suppose that the point $q \in \Psi_p$ is a point of local compactness of X. Then there exists a neighborhood U of the point q whose closure $\bar{U}$

is compact, and a number $T > 0$ such that whatever the point $r \in f(p, R^+)$, we have that

$$f(r, [0, T]) \cap U \neq \phi.$$

Therefore there exists a $\tau \in [0, T]$ such that $f(r, \tau) \in U$. This implies the inclusion

$$r \in f(U, -\tau) \subseteq f(U, [-T, 0])$$

and, since r is arbitrary, the semitrajectory

$$f(p, R^+) \subseteq f(U, [-T, 0]) \subseteq f(\bar{U}, [-T, 0]). \tag{3.25}$$

It follows from (3.25), by virtue of the compactness of the set $f(\bar{U}, [-T, 0])$ that the motion $f(p, t)$ is positively Lagrange stable. This completes the proof of the lemma.

Proof of Theorem 3.25. Suppose Ω_p contains two distinct minimal sets M_1 and M_2. Then, according to Lemma 2.25, either $\Psi_p = M_1$ or $\Psi_p = \phi$. The equality $\Psi_p = M_1$ is impossible since, by the same lemma, either $\Psi_p = M_2$ or $\Psi_p = \phi$. Thus $\Psi_p = \phi$.

Now assume that the space X is locally compact and that the motion $f(p, t)$ is not positively Lagrange stable. In this case all the points of the space X are points of local compactness and on the basis of Lemma 3.25 the set $\Psi_p = \phi$.

We now establish two corollaries which follow from Theorem 3.25.

COROLLARY 2.25. *In a locally compact space the set Ψ_p is compact.*

In fact, if X is locally compact and $\Psi_p \neq \phi$ then the motion $f(p, t)$ must be positively Lagrange stable. Then Ω_p is compact and Ψ_p, being a closed subset of Ω_p, is also compact.

As was established by M. V. Bebutov [2], in a compact space almost recurrence reduces to recurrence. However, the following more general proposition holds.

COROLLARY 3.25. *In a locally compact space every almost recurrent motion is recurrent.*

In fact, if the motion $f(p, t)$ is almost recurrent, then $p \in \Psi_p \cap B_p$ and consequently $\Psi_p \cap B_p \neq \phi$. Furthermore, on the basis of Theorem 3.25, the motion $f(p, t)$ is (both positively and negatively) Lagrange stable. It follows from the last fact, the almost recurrence of the motion $f(p, t)$, and Theorem 4.18 that $f(p, t)$ is recurrent.

It is well known (Corollary 2.16) that the positive Lagrange stability of a motion $f(p, t)$ guarantees the existence of at least one minimal set in Ω_p.

THEOREM 4.25. *If the motion $f(p, t)$ is positively Lagrange stable and M is the only minimal set in Ω_p, then $\Psi_p = M$ and the semitrajectory $f(p, R^+)$ uniformly approximates M.*

Proof. Assume that the motion $f(p, t)$ is positively Lagrange stable and that M is the only minimal set in Ω_p. Suppose the semitrajectory $f(p, R^+)$ does not approximate M uniformly. Then there exists an $\varepsilon_0 > 0$ such that for any number $T_n > 0$ whatever, there exists a pair of points $p_n \in f(p, R^+)$ and $q_n \in M$ for which

$$\rho(f(p_n, [0, T_n]), q_n) \geq \varepsilon_0. \tag{4.25}$$

We choose the numbers $T_n \geq 0$ $(n = 1, 2, \ldots)$ so that $\{T_n\} \to +\infty$. Since $p_n \in \Sigma_p^+$ and $q_n \in \Omega_p \subseteq \Sigma_p^+$ and Σ_p^+ is compact, the sequences $\{p_n\}$ and $\{q_n\}$ can be assumed to be convergent. Let $\{p_n\} \to p_0$ and $\{q_n\} \to q_0$. Then $p_0 \in \Sigma_p^+$, $q_0 \in M$, and from (4.25) it follows that

$$\rho(f(p_0, R^+), q_0) \geq \varepsilon_0. \tag{5.25}$$

But $\Omega_{p_0} \subseteq \Omega_p$ (since $\Sigma_p^+ = f(p, R^+) \cup \Omega_p$), and therefore Ω_{p_0} is compact and contains a minimal set. Inasmuch as M is the only minimal set in Ω_p, then $M \subseteq \Omega_{p_0}$ and, in particular, $q_0 \in \Omega_{p_0}$. This contradicts inequality (5.25).

From this we conclude that the semitrajectory $f(p, R^+)$ uniformly approximates M. Then $M \subseteq \Psi_p$ and $\Psi_p \neq \phi$. On the basis of Lemma 2.25 the set $\Psi_p = M$. This completes the proof of the theorem.

Note that in the case under consideration the set M is a compact minimal set and therefore all motions in Ψ_p are recurrent.

Remark 1.25. In locally compact spaces there exist examples in which Ω_p contains only one minimal set and $\Psi_p = \phi$. To construct such an example it suffices to consider the case when X is locally compact and Ω_p contains only two minimal sets and then remove one of these sets. In such a case, however, Ω_p will not be compact.

Since a necessary and sufficient condition for a motion $f(p, t)$ to be positively Lagrange stable in a locally compact space is that Ω_p be non-empty and compact (see Theorem 2.8), then Theorems 3.25 and 4.25 imply the following corollary.

COROLLARY 4.25.   *In a locally compact space, $\Psi_p \neq \phi$ if and only if Ω_p is compact and contains only one minimal set M. Furthermore, $\Psi_p = M$.*

THEOREM 5.25.   *A necessary and sufficient condition for the semitrajectory $f(p, R^+)$ of a positively Lagrange stable motion $f(p, t)$ to uniformly approximate a subset $Q \subseteq \Omega_p$ is that Σ_Q be the only minimal set in Ω_p.*

Proof. Suppose $f(p, R^+)$ uniformly approximates some subset $Q \subseteq \Omega_p$. Then $Q \subseteq \Psi_p$ and $\Psi_p \neq \phi$. According to Theorems 3.25 and 4.25, the set Ψ_p is the only minimal set in Ω_p. Furthermore, $\Sigma_Q = \Psi_p$. Conversely, let Σ_Q be the only minimal set in Ω_p. Then, by Theorem 4.25, $\Psi_p = \Sigma_Q$, and the semitrajectory $f(p, R^+)$ uniformly approximates Ψ_p and *a fortiori* its subset Q.

COROLLARY 5.25.   *The set Ω_p of all ω-limit points of the positively Lagrange stable motion $f(p, t)$ is a minimal set (of recurrent motions) if and only if the semitrajectory $f(p, R^+)$ uniformly approximates Ω_p.*

[This proposition was first established by V. V. Nemytsky [1].] Motions whose ω-limit set contains only one minimal set were first considered by G. D. Birkhoff [1] under the name positively semi-asymptotic motions.

Remark 2.25.   On the real line R^1 we always have $\Psi_p = \Omega_p$ since Ω_p is either empty or consists of one point, and then the motion $f(p, t)$ is positively Lagrange stable.

The following proposition is a generalization of a theorem of V. V. Nemytsky [1]. It was established in works of the author [2] and also of L. Deysach and G. Sell [1].

THEOREM 6.25.   *If the motion $f(p, t)$ is positively Lagrange stable and $f(p, R^+)$ is positively Lyapunov stable relative to $f(p, R^+)$, then Ω_p is a minimal set of almost periodic motions.*

Proof. First note that the positive Lagrange stability of the motion $f(p, t)$ guarantees the compactness of the set Ω_p and the Lagrange stability of all motions in it. According to Theorem 3.22, the set Σ_p^+ is uniformly positively Lyapunov stable relative to Σ_p^+. Inasmuch as $\Omega_p \subseteq \Sigma_p^+$, then Ω_p is also uniformly positively Lyapunov stable relative to Ω_p. On the basis of Theorem 3.24, all the motions in Ω_p are almost periodic. It remains to prove that the following lemma holds.

LEMMA 4.25. *If $f(p, R^+)$ is uniformly positively Lyapunov stable relative to $f(p, R^+)$ and $\Omega_p \neq \phi$, then Ω_p is a minimal set.*

Proof. Suppose that the set Ω_p is not minimal. Then there exists a point $q \in \Omega_p$ such that $\Sigma_q \neq \Omega_p$. Let $r \in \Omega_p$ but $r \notin \Omega_q$. By virtue of the fact that Σ_q is closed, the distance $\rho(r, \Sigma_q) = d > 0$. Since the set $f(p, R^+)$ is uniformly positively Lyapunov stable relative to $f(p, R^+)$, there exists a number $\delta > 0$ corresponding to $d/3$. Consider the neighborhoods $S(q, \delta/2)$ and $S(r, d/3)$. There exist positive numbers t_1 and t_2 such that $T \equiv t_2 - t_1 > 0$,

$$f(p, t_1) \in S\left(q, \frac{\delta}{2}\right), \ f(p, t_2) \in S\left(r, \frac{d}{3}\right).$$

Then, provided $f(p, t) \in S(q, \delta/2)$ for positive t,

$$\rho(f(p, t), f(p, t_1)) < \delta$$

and, by virtue of the choice of δ,

$$\rho(f(p, t+T), f(p, t_2)) < \frac{d}{3},$$

i.e.,

$$f(p, t+T) \in S\left(f(p, t_2), \frac{d}{3}\right) \subseteq S\left(r, \frac{2d}{3}\right).$$

Therefore

$$f(p, t+T) \notin S\left(\Sigma_q, \frac{d}{3}\right).$$

This contradicts the fact that, since $f(p, t)$ is continuous, for the numbers T and $d/3$ and point q there exists a $\delta_1 < \delta/2$ such that $\rho(f(q_1, T), f(q, T)) <$

$d/3$ provided $\rho(q_1, q) < \delta_1$. This completes the proof of the lemma.

Theorem 6.25 implies the following corollary.

COROLLARY 6.25.   *If the motion $f(p, t)$ is both positively Lagrange stable and positively Poisson stable and $f(p, R^+)$ is uniformly positively Lyapunov stable relative to $f(p, R^+)$, then $f(p, t)$ is almost periodic.*

THEOREM 7.25.   *A necessary and sufficient condition for the ω-limit set of a positively Lagrange stable motion $f(p, t)$ to be the trajectory of a singular motion is that there exist for each $q \in \Omega_p$ a sequence $\{t_n\}$ which is relatively dense on R^+ such that $\{f(p, t_n)\} \to q$.*

We shall first show that the following lemma holds.

LEMMA 5.25.   *If there exists a relatively dense sequence $\{t_n\}$ on R^+ such that $\{f(p, t_n)\} \to q$, then the motion $f(q, t)$ is singular.*

Proof. Suppose there exist $L > 0$ and a sequence $\{t_n\}$ such that for any $l > 0$ there occurs at least one point of the sequence $\{t_n\}$ on the segment $[l, l+L)$ and suppose also that $\{f(p, t_n)\} \to q$. Without loss of generality, we may assume that $L \leqq t_{n+1} - t_n \leqq 3L$ for all n. To this end, it suffices to replace the sequence $\{t_n\}$ by a subsequence of it by choosing one term from $\{t_n\}$ on each segment of the form

$$[2kL, (2k+1)L] \quad (k = 0, 1, 2, \ldots).$$

Let $\tau_n = t_{n+1} - t_n$. Since the sequence $\{\tau_n\}$ is bounded, it contains a convergent subsequence, say $\{\tau_{n_k}\} = \{t_{n_k+1} - t_{n_k}\}$. Let $\lim_{k \to \infty} \tau_{n_k} = \tau$. Then $\tau \geqq L > 0$ and

$$\{f(p, t_{n_k} + \tau_{n_k})\} = \{f(f(p, t_{n_k}), \tau_{n_k})\} \to f(q, \tau).$$

On the other hand,

$$\{f(p, t_{n_k} + \tau_{n_k})\} = \{f(p, t_{n_k+1})\} \to q.$$

It follows that $f(q, \tau) = q$ where $\tau > 0$ and consequently the motion $f(q, t)$ is singular.

The analogous proposition for the special case of a dynamical system was established by M. I. Almukhamedov [1].

Proof of Theorem 7.25. Assume that the motion $f(p, t)$ is positively Lagrange stable and that Ω_p is the trajectory of a singular motion. Take an arbitrary point $q \in \Omega_p$. Since the motion $f(q, t)$ is singular, there exists a $\tau > 0$ such that

$$f(q, t+\tau) = f(q, t)$$

for all t. Consider the arc

$$f(p, [n\tau, (n+1)\tau]) \tag{6.25}$$

of the semitrajectory $f(p, R^+)$. On $[n\tau, (n+1)\tau]$ there exists a number t_n such that the point $f(p, t_n)$ on the arc (6.25) is closest to q. Clearly, the sequence $\{t_n\}$ is relatively dense on R^+. We shall prove that $\{f(p, t_n)\} \to q$. Since Σ_p^+ is compact, the sequence $\{f(p, t_n)\}$ has at least one limit point q'. Now choose a subsequence $\{f(p, t_{n_k})\} \to q'$ such that $\{f(p, n_k\tau)\}$ converges. Let $\{f(p, n_k\tau)\} \to q''$. Then there exists a number $\tau'' \in [0, \tau]$ such that $f(q'', \tau'') = q$. Furthermore,

$$\{f(p, n_k\tau+\tau'')\} = \{f(f(p, n_k\tau), \tau'')\} \to f(q'', \tau'') = q.$$

Since, according to the choice of the numbers t_{n_k}, the inequality

$$\rho(f(p, t_{n_k}), q) \leqq \rho(f(p, n_k\tau+\tau''), q)$$

holds, we also have that $\{f(p, t_{n_k})\} \to q$. Hence $q' = q$ and then q is the only limit point of the sequence $\{f(p, t_n)\}$; therefore $\{f(p, t_n)\} \to q$.

Now assume that for every point $q \in \Omega_p$ there exists a relatively dense sequence $\{t_n\}$ on R^+ such that $\{f(p, t_n)\} \to q$. Then $\Psi_p = \Omega_p \neq \phi$. On the basis of Lemma 5.25, all motions in Ω_p are singular and, on the basis of Theorem 3.25, the set Ω_p can contain at most one minimal set. Therefore Ω_p must be the trajectory of one singular motion. This completes the proof of the theorem.

Note that in proving sufficiency, we did not use the condition that the motion $f(p, t)$ is positively Lagrange stable.

§ 26   Stability of rest points

If the point p is a rest point then the following definition agrees with Definition 1.22.

Definition 1.26. A rest point p is said to be *positively* (*negatively*) *Lyapunov stable* if for any $\varepsilon > 0$ there exists a $\delta > 0$ such that

$$f(q, t) \in S(p, \varepsilon)$$

provided $q \in S(p, \delta)$ and $t \in R^+$ ($t \in R^-$). If, moreover,

$$\rho(f(q, t), p) \to 0 \qquad (1.26)$$

as $t \to +\infty$ ($t \to -\infty$) then the rest point p is said to be *positively* (*negatively*) *asymptotically Lyapunov stable.*

Definition 2.26. A nonnegative real-valued function $V(q)$, defined on a neighborhood $S(p, \delta_0)$ of a rest point p, is called a *Lyapunov function for the point* p if the following two conditions are satisfied:

1.26. $\{V(q_n)\} \to 0$ if and only if $\{q_n\} \to p$;

2.26. If $t > 0$ and $f(q, [0, t]) \subseteq S(p, \delta_0)$, then $V(f(q, t)) \leq V(q)$.

THEOREM 1.26.   *A necessary and sufficient condition for a rest point p to be positively Lyapunov stable is that there exist a Lyapunov function for p.*

Proof. Necessity. Let the point p be positively Lyapunov stable and $\varepsilon_0 > 0$. According to Definition 1.26, there exists a $\delta_0 > 0$ such that when $\rho(p, q) < \delta_0$ then $\rho(f(q, t), p) < \varepsilon_0$ for all $t \geq 0$. Set

$$V(q) = \sup_{t \geq 0} \rho(f(q, t), p). \qquad (2.26)$$

Thus the function $V(q)$ is defined in $S(p, \delta_0)$. We shall prove that condition 1.26 is satisfied. Let $\{q_n\} \to p$ and $\varepsilon > 0$. There exists a $\delta > 0$ such that, for $\rho(q, p) < \delta$,

$$f(q, R^+) \subseteq S(p, \varepsilon),$$

and an N such that, for $n > N$, $\rho(q_n, p) < \delta$. Then

$$f(q_n, R^+) \subseteq S(p, \varepsilon).$$

Furthermore,

$$\rho(f(q_n, t), p) < \varepsilon$$

for all $t \geq 0$ and then

$$V(q_n) \leqq \varepsilon \text{ for all } n > N,$$

i.e., $\{V(q_n)\} \to 0$.

Assume that $\{V(q_n)\} \to 0$. Then it follows from (2.26) that $\rho(q_n, p) \to 0$ and hence $\{q_n\} \to p$.

To verify condition 2.26 assume that $t > 0$ and

$$f(q, [0, t]) \subseteq S(p, \delta_0).$$

Then

$$V(f(q, t)) = \sup_{\tau \geq 0} \rho(f(q, t+\tau), p) \leqq \sup_{\tau \geq 0} \rho(f(q, \tau), p) = V(q).$$

Sufficiency. Let $V(q)$ be a Lyapunov function for the rest point p. Take $\varepsilon > 0 (\varepsilon < \delta_0)$ and set

$$\lambda = \inf_{\rho(q, p) = \varepsilon} V(q).$$

By condition 1.26, the number $\lambda > 0$ and there exists a $\delta > 0$ such that the inequality $V(q) < \lambda$ is satisfied when $\rho(q, p) < \delta$. We shall show that this δ is a suitable choice for the δ of Definition 1.26. Suppose the contrary, i.e., assume that there exists a point $q \in S(p, \delta)$ such that

$$f(q, R^+) \cap (X \setminus S(p, \varepsilon)) \neq \phi.$$

Denote by τ the first moment of departure of the point q from $S(p, \varepsilon)$. Then

$$\rho(f(q, \tau), p) = \varepsilon \quad \text{and} \quad V(f(q, \tau)) \geqq \lambda.$$

But, by virtue of 2.26,

$$V(f(q, \tau)) \leqq V(q) < \lambda.$$

The contradiction thus obtained shows that the assumption made above is not true.

THEOREM 2.26.   *A necessary and sufficient condition for a rest point p to be asymptotically positively Lyapunov stable is that there exist a Lyapunov function $V(q)$ for the point p such that*

$$V(f(q, t)) \to 0 \quad as \quad t \to +\infty \tag{3.26}$$

provided

$$f(q, R^+) \subseteq S(p, \delta_0).$$

Proof. Necessity. Let the rest point p be asymptotically positively Lyapunov stable. By Theorem 1.26 there exists a Lyapunov function $V(q)$ for the point p. Furthermore, if $f(q, R^+) \subseteq S(p, \delta_0)$, then condition (1.26) is satisfied and then according to property 1.26 (of a Lyapunov function) condition (3.26) holds.

Sufficiency. Suppose there exists a Lyapunov function $V(q)$ possessing property (3.26) for the rest point p. Then, by Theorem 1.26, the point p is positively Lyapunov stable, i.e., for any $\varepsilon > 0$ there exists a $\delta > 0$ such that $\rho(f(q, t), p) < \varepsilon$ provided $\rho(q, p) < \delta$ and $t \geq 0$. If this δ has been found for $\min(\varepsilon, \delta_0)$, then, for any point $q \in S(p, \delta)$,

$$f(q, R^+) \subseteq S(p, \delta_0),$$

and therefore condition (3.26) is satisfied. Furthermore, according to property 1.26 (of a Lyapunov function) condition (1.26) must be satisfied and consequently the point is asymptotically positively Lyapunov stable. This completes the proof of Theorem 2.26.

LEMMA 1.26.   *If there exists a point $q \neq p$ such that $p \in A_q (p \in \Omega_q)$, then the rest point p is not positively (negatively) Lyapunov stable.*

Proof. Let, for instance, $q \neq p$, $p \in A_q$ and $\varepsilon_0 = \rho(p, q)$. Since $p \in A_q$, there exists a sequence of negative numbers $\{t_n\} \to -\infty$ such that

$$\{q_n\} = \{f(q, t_n)\} \to p.$$

Furthermore, for any $\delta > 0$ there exists a point $q_n \in S(p, \delta)$ and for this point

$$f(q_n, -t_n) = q \notin S(p, \varepsilon_0)(-t_n > 0).$$

123

Thus, in any neighborhood of the point p there are points which for $t > 0$ depart from $S(p, \varepsilon_0)$ and consequently the point p is not positively Lyapunov stable.

Definition 3.26.   A rest point $p \in X$ is said to be a *simple* rest point if every neighborhood of p contains at least one point which is different from p and there exists a neighborhood $S(p, \varepsilon_0)$ of the point p with compact closure which does not contain any dynamical limit points other than p.

LEMMA 2.26.   *A simple rest point is not Lyapunov stable.*

Proof. If there exists a point $q \neq p$ such that $p \in \Delta_q$ then, by Lemma 1.26, the point p is either not negatively Lyapunov stable or not positively Lyapunov stable (or both). Otherwise, for all points q in the neighborhood $S(p, \varepsilon_0) \setminus p$,

$$\Delta_q \cap S(p, \varepsilon_0) = \phi,$$

and then the semitrajectory $f(q, R^-)$ $[f(q, R^+)]$ departs from $S(p, \varepsilon_0)$ and the point p is not Lyapunov stable. This completes the proof of the lemma.

It is easy to see that if p is a simple rest point and $p \in \Omega_q$, then $\Omega_q = \{p\}$. In fact, if Ω_q contained, besides p, a point $r \neq p$, then $r \notin S(p, \varepsilon_0)$ and therefore $\rho(r, p) > \varepsilon_0/2$. Furthermore, there would be a point of Ω_q at a distance $\varepsilon_0/2$ from p, which is impossible. In this connection, it is possible to assert that if p is a simple rest point, then any point $q \in S(p, \varepsilon_0)$ different from p will be of one of the following three types:

(1) *p-hyperbolic* if $p \notin \Delta_q \cup \Omega_q$;
(2) *p-parabolic* if $\Omega_q = \{p\}$, $p \notin A_q$ or $A_q = \{p\}$, $p \notin \Omega_q$;
(3) *p-elliptic* if $\Omega_q = A_q = \{p\}$.

Note that if the point q is *p*-hyperbolic, then any point $q' \in f(q, R)$ $\cap S(p, \varepsilon_0)$ will also be *p*-hyperbolic. An analogous assertion also holds for points of the other two types.

Starting from this classification of points and trajectories of a neighborhood of a point p, the following simple rest points p are logically possible:

(1) *hyperbolic* if there exists an $\varepsilon > 0$ such that every point $q \in S(p, \varepsilon) \setminus p$ is *p*-hyperbolic;

(2) *elliptic* if there exists an $\varepsilon > 0$ such that every point $q \in S(p, \varepsilon) \setminus p$ is *p*-elliptic;

(3) *parabolic* if there exists an $\varepsilon > 0$ such that every point $q \in S(p, \varepsilon) \setminus p$ is *p*-parabolic;

(4) *hyperbolico-elliptic* if there exists an $\varepsilon > 0$ such that the neighborhood $S(p, \varepsilon) \setminus p$ contains no p-parabolic points and every neighborhood of the point p contains p-hyperbolic as well as p-elliptic points;

(5) *hyperbolico-parabolic* if there exists an $\varepsilon > 0$ such that the neighborhood $S(p, \varepsilon) \setminus p$ contains no p-elliptic points and every neighborhood of the point p contains p-hyperbolic as well as p-parabolic points;

(6) *elliptico-parabolic* if there exists an $\varepsilon > 0$ such that the neighborhood $S(p, \varepsilon) \setminus p$ contains no p-hyperbolic points and every neighborhood of the point p contains p-elliptic as well as p-parabolic points;

(7) *mixed type* if every neighborhood of the point p contains points of all three types.

We shall prove that there do not exist hyperbolic or hyperbolico-elliptic simple rest points, i.e., the following theorem is valid.

THEOREM 3.26 (N. N. LADIS [2]). *If p is a simple rest point in any neighborhood of which there exist p-hyperbolic points, then every neighborhood of p contains p-parabolic points.*

Before proving this theorem we shall prove the following lemma.

LEMMA 3.26. *If V is a neighborhood of a simple rest point and $\bar{V}$ is compact, then $f(V, R)$ is a locally compact space.*

Proof. Set $Q \equiv f(V, R)$ and let $q \in Q$, i.e., there exists a point $r \in V$ and a moment $t_0 \in R$ such that $q = f(r, t_0)$. Then $r = f(q, -t_0)$. Since the function f is continuous, there exists a neighborhood U of the point q such that $f(\bar{U}, -t_0) \subseteq V$. Furthermore, $\bar{U} \subseteq Q$ and $f(\bar{U}, -t_0)$, being a closed subset of the compactum $\bar{V}$, is itself compact. And then

$$\bar{U} = f(f(\bar{U}, -t_0), t_0)$$

is also compact. Thus, Q is a locally compact space, which is what was required to be proved.

Proof of Theorem 3.26. Since p is a simple rest point, there exists a neighborhood $V \equiv S(p, \delta_0)$ of p such that its closure is compact and does not contain other dynamical limit points except p. Consider the set $Q \equiv f(V, R)$. By Lemma 3.26, it is a locally compact space.

According to P. S. Aleksandrov's theorem one can consider the one point compactification $Q' = Q \cup \xi$ of the space Q, declaring ξ to be a rest point: $f(\xi, t) = \xi$ for any $t \in R$. [See Theorem 9 in P. S. Aleksandrov [2], page 390: To any locally compact space Q one can adjoin one point ξ so that a compactum $Q' = Q \cup \xi$ is obtained (and furthermore the topology in Q as a set contained in the compactum Q' will coincide with the topology given *a priori* in Q).] The space Q' can be metrized, a fact we shall assume.

In the dynamical system f on Q' all motions are Lagrange stable and the only dynamical limit points will be p and ξ. No other point q of Q can be a dynamical limit point since, by virtue of the invariance of the dynamical limit sets, all points of the trajectory $f(q, R)$ would be dynamical limit points. But then, in view of the fact that

$$(V \setminus p) \cap f(q, R) \neq \phi,$$

there would be dynamical limit points in the set $V \setminus p$ which contradicts our initial assumption. In view of what has been said, and using Theorem 3.8 on the connectivity of limit sets, we conclude that if $q \in Q$, then either $A_q = \Omega_q = \{p\}$ or $A_q = \Omega_q = \{\xi\}$ or $A_q = \{p\}, \Omega_q = \{\xi\}$ or $A_q = \{\xi\}, \Omega_q = \{p\}$. Let

$$V_1 \supset V_2 \supset V_3 \supset \ldots \supset V_n \supset \ldots \ni p$$

and

$$U_1 \supset U_2 \supset U_3 \supset \ldots \supset U_n \supset \ldots \ni \xi$$

be nested sequences of neighborhoods which contract to the points p and ξ, respectively, for which $\bar{V}_1 \subset V$ and $\bar{U}_1 \cap \bar{V} = \phi$.

According to our assumption, there exists a p-hyperbolic point x_n in V_n. Then $A_{x_n} = \Omega_{x_n} = \{\xi\}$.

We shall say that the arc $f(r, [t', t''])$ joins the neighborhood V_n with the neighborhood U_n if the points $f(r, t')$ and $f(r, t'')$ belong to the boundaries of V_n and U_n respectively, $t' < t''$, and

$$f(r, (t', t'')) \cap (\bar{V}_n \cup \bar{U}_n) = \phi.$$

Let $f(x_n, [t_n, t_n'])$ be an arc of the semitrajectory $f(x_n, R^+)$ which joins V_n with U_n. On it is an arc $f(x_n, [t_{n1}, t_{n1}'])$ which joins V_{n-1} with U_{n-1}, on this arc there is an arc $f(x_n, [t_{n2}, t_{n2}'])$ which joins V_{n-2} with U_{n-2}, and so forth. On

$$f(x_n, [t_{n(n-2)}, t_{n(n-2)}'])$$

there is an arc

$$f(x_n, [t_{n(n-1)}, t'_{n(n-1)}])$$

which joins V_1 with U_1. Let $q_n = f(x_n, t_{n(n-1)})$. The sequence $\{q_n\}$ can be considered to be convergent. Let $\{q_n\} \to q$. We shall prove that the point q is p-parabolic in the initial dynamical system, namely that $\{p\} = A_q$ and $p \notin \Omega_q$. For the dynamical system on Q' this is equivalent to asserting that $A_q = \{p\}$, $\Omega_q = \{\xi\}$.

Fix an arbitrary number n_0 and for any $n > n_0$ denote by τ_n and τ'_n positive numbers such that the arc $f(q_n, [-\tau_n, \tau'_n])$ joins V_{n_0} with U_{n_0}.

If for any natural number n there existed in the set

$$M \equiv Q' \setminus (\bar{V}_{n_0} \cup \bar{U}_{n_0})$$

an arc $f(r_n, [0, n])$ of time length n, then, denoting a limit point for $\{r_n\}$ by r_0, we obtain that $f(r_0, R^+) \subseteq \bar{M}$, which is impossible since there are no dynamical limit points in $\bar{M}$. Therefore there exists a number $T > 0$ such that there are no arcs (of trajectories) of time length T in M.

According to the choice of the points q_n it is necessary that

$$\tau_n < T \text{ and } \tau'_n < T.$$

Thus the sequences $\{\tau_n\}$ and $\{\tau'_n\}$ are bounded. Let them have limit points τ_0 and τ'_0 respectively so that

$$\tau_0 = \lim_{k \to \infty} \tau_{n_k}, \tau'_0 = \lim_{k \to \infty} \tau'_{n_k}.$$

Then

$$\{f(q_{n_k}, -\tau_{n_k})\} \to f(q, -\tau_0), \{f(q_{n_k}, \tau'_{n_k})\} \to f(q, \tau'_0)$$

and the points $f(q, -\tau_0)$ and $f(q, \tau'_0)$ belong to the boundaries of V_{n_0} and U_{n_0} respectively. Consequently,

$$f(q, R^-) \cap \bar{V}_{n_0} \neq \phi, f(q, R^+) \cap \bar{U}_{n_0} \neq \phi$$

for any n_0. Therefore $p \in \Sigma_p^-$, $\xi \in \Sigma_q^+$ from which it follows that in Q' we have that $\{p\} = A_q$, $\{\xi\} = \Omega_q$, which is what was required to be proved.

We have thus proved the absence of hyperbolic and hyperbolico-elliptic simple rest points.

The existence of simple rest points of the remaining five types is easily verified by means of examples.

In particular, one can consider the one rest point on the circle as an elliptic point. N. N. Ladis [1] showed that in a connected locally compact, but not compact, space there are no simple elliptic rest points.

We now proceed to the consideration of Lyapunov stable rest points.

We shall call a Lyapunov stable rest point p a point of "center" type if there exists an $\varepsilon > 0$ such that for any point $q \in S(p, \varepsilon) \setminus p$ the closure of the trajectory $f(q, R)$ is a minimal set.

We shall call a Lyapunov stable rest point p a point of "centro-focus" type if it is not a point of "center" type but every neighborhood of it contains a point distinct from p the closure of whose trajectory is a minimal set.

THEOREM 4.26.   *If a Lyapunov stable rest point is a point of local compactness but is not a point of "center" type, then it is a point of "centro-focus" type.*

Proof. Suppose the conditions of the theorem are satisfied. Then, for any $\varepsilon > 0$, there exists a $\delta > 0$ such that

$$f(S(p, \delta), R) \subseteq S(p, \varepsilon).$$

Since p is a point of local compactness, there exists an $\varepsilon_0 > 0$ such that $\bar{S}(p, \varepsilon_0)$ is a compactum. Now let $\varepsilon < \varepsilon_0$. Then, since $f(r, R) \subseteq S(p, \varepsilon)$ for any point $r \in S(p, \delta) \setminus p$, we have that $\Sigma_r \subseteq \bar{S}(p, \varepsilon)$ and is a compactum. Therefore Σ_r contains some minimal subset Σ. According to Lemma 1.26, $p \notin \Sigma_r$. Then $p \notin \Sigma$, and since for every point $q \in \Sigma$ we have $\Sigma_q = \Sigma$, then in any neighborhood of the point p there exists a point q, different from p, the closure of whose trajectory is a minimal set. Consequently, the point p is a point of "centro-focus" type, which is what was required to be proved.

The classification of rest points introduced here is a generalization to the case of arbitrary metric spaces of Nemytsky's classification (see N. N. Nemytsky [7]).

In concluding this section we will present one more theorem, whose proof is due to M. S. Izman. We shall first show that the following lemma holds.

Lemma 4.26.   *Suppose the rest point p which has a compact neighborhood is not positively (negatively) Lyapunov stable. Then every neighborhood $S(p, \varepsilon) \setminus p$ of p contains negative (positive) semitrajectories.*

Proof. Suppose the point p is not positively Lyapunov stable. Then there exists an $\varepsilon_0 > 0$ such that, for any $\delta > 0$,

$$f(q, R^+) \cap (X \setminus S(p, \varepsilon_0)) \neq \phi$$

for some point $q \in S(p, \delta)$. We can assume that $S(p, \varepsilon_0)$ is compact and that $\varepsilon \in (0, \varepsilon_0)$. There thus exist a sequence of points $\{p_n\}$, $\{p_n\} \to p$, and a sequence of moments of time $\{t_n\}$, $t_n \geq 0$, such that

$$f(p_n, [0, t_n)) \subseteq S\left(p, \frac{\varepsilon}{2}\right), \; \rho(f(p_n, t_n), p) = \frac{\varepsilon}{2}. \tag{4.26}$$

Let $q_n = f(p_n, t_n)$. We can assume that $\{q_n\} \to r$ and $\{t_n\} \to +\infty$ since if the sequence $\{t_n\}$ had a finite limit point t_0, then $r = f(p, t_0)$, which is impossible inasmuch as $\rho(r, p) = \varepsilon/2$, and p is a rest point.

We shall prove that $f(r, R^-) \subseteq S(p, \varepsilon)$. Suppose the contrary. Then there exists a $\tau < 0$ such that $\rho(f(r, \tau), p) = \varepsilon$. Since $\{f(q_n, \tau)\} \to f(r, \tau)$ and $\{t_n\} \to +\infty$, there exists an index N such that $\rho(f(q_N, \tau), p) > \varepsilon/2, t_N > -\tau$. Then

$$\rho(f(p_N, t_N + \tau), p) > \frac{\varepsilon}{2}, 0 < t_N + \tau < t_N.$$

But this contradicts (4.26). Consequently, $f(r, R^-) \subseteq S(p, \varepsilon)$. This completes the proof of the lemma.

Theorem 5.26.   *A necessary and sufficient condition for a rest point p which possesses a compact neighborhood to be asymptotically positively (negatively) Lyapunov stable is that there exist a neighborhood of this point $S(p, \varepsilon_0) \setminus p$ which does not contain negative (positive) semitrajectories.*

Proof. Necessity. Suppose the point p is asymptotically positively Lyapunov stable and that $\varepsilon > 0$ is such that the neighborhood $S(p, \varepsilon)$ has compact closure. For the number $\varepsilon/2$ there exists a $\delta > 0 (\delta < \varepsilon)$ such that, whatever the point $q \in S(p, \delta)$ is, we have

$$f(q, R^+) \subseteq S(p, \varepsilon)$$

and condition (1.26) is satisfied as $t \to +\infty$. We shall show, by assuming the contrary and arriving at a contradiction, that the neighborhood $S(p, \delta/2)\setminus p$ does not contain negative semitrajectories.

Let $f(q, R^-) \subset S(p, \delta/2)\setminus p$. Then $A_q \neq \phi$ and $A_q \subset S(p, \delta)$. Since A_q is invariant and closed and all points in $S(p, \delta)$ (and including the points of the set A_q), by condition (1.26), tend to p as $t \to +\infty$, then $p \in A_q$. Then, by Lemma 1.26, the point p is not positively Lyapunov stable, which contradicts our assumption.

Sufficiency. Suppose there exists an $\varepsilon_0 > 0$ such that $S(p, \varepsilon_0)\setminus p$ does not contain negative semitrajectories. Then, according to Lemma 4.26, the point p is positively Lyapunov stable. By the definition of positive Lyapunov stability we choose a corresponding $\delta = \delta(\varepsilon_0/2) > 0$. From the fact that p possesses a compact neighborhood and in agreement with the choice of δ it follows that for any point $q \in S(p, \delta)\setminus p$ the set $\Omega_q \neq \phi$ and $\Omega_q \subset S(p, \varepsilon_0)$. Since Ω_q is invariant and $S(p, \varepsilon_0)\setminus p$ does not contain negative semitrajectories, then $\Omega_q = \{p\}$. Therefore, for any point $q \in S(p, \delta)\setminus p$, condition (1.26) is satisfied and the point p is positively Lyapunov stable. This completes the proof of the theorem.

All the definitions and results of this section carry over easily to the case where instead of a rest point p one considers an arbitrary closed invariant set M. See, for example, the work of V. I. Zubov [1], N. N. Krasovsky [1], J. Auslander and P. Seibert [1, 2], N. P. Bhatia and G. P. Szegö [1], L. A. Chelysheva [1], and M. S. Izman [1, 2].

Chapter VI

GENERALIZED THEORY OF DYNAMICAL SYSTEMS

§ 27 General dynamical systems. Topological transformation groups

We now consider an arbitrary topological space X and an arbitrary group G. [Concerning topological spaces, see, e.g., P. S. Aleksandrov [2].] We shall assume that each point $p \in X$ and each element $g \in G$ is assigned a point $q = f(p, g) \in X$ so that:

1.27. $f(p, 0) = p$ for any point $p \in X$ (0 is the zero of the group G).

2.27. The function $f(p, g)$ is continuous with respect to p, i.e., for any neighborhood W of the point $f(p_0, g_0)$ there exists a neighborhood U of the point p_0 such that

$$f(U, g_0) \subseteq W.$$

3.27. $f(f(p, g_1), g_2) = f(p, g_1 + g_2)$ for any point $p \in X$ and arbitrary $g_1, g_2 \in G$.

If this is the situation we say that a general dynamical system $[X, G, f]$ has been defined.

The set of points

$$f(p, G) = \bigcup_{g \in G} f(p, g)$$

is called the *orbit of the point p*.

The general dynamical system $[X, G, f]$ is said to be a *continuous general dynamical system* (or a *topological transformation group*) if G is a topological group (see, e.g., L. S. Pontryagin [1]), and the function $f(p, g)$ is continuous in both arguments simultaneously, i.e., for any neighborhood W of the point $f(p_0, g_0)$ there exist neighborhoods U and V of the point p_0 and of the element g_0 respectively such that

$$f(U, V) \subseteq W.$$

Clearly, the dynamical systems $[X, R, f]$ considered in the preceding chapters are special cases of continuous generalized dynamical systems.

Example 1.27.   Let $X = R^3$ and let G be the group of vectors in two-dimensional space with respect to the operation of vector addition. Every vector is determined by its projections on the x- and y-coordinate axes. Take a point $p(x_0, y_0, z_0) \in R^3$ and a vector $g(X_0, Y_0) \in G$. Set the point $q(x_0 + X_0, y_0 + Y_0, z_0) \in R^3$ into correspondence with the point $p(x_0, y_0, z_0)$. Thus, the function $q = f(p, g)$ which has been defined satisfies all the requirements for a general dynamical system. The orbits are planes which are parallel to the xy-plane.

Continuous general dynamical systems were first introduced and studied by V. V. Nemytsky [2, 3, 5], E. A. Barbashin [4, 7] and several American mathematicians (see, e.g., the book by W. H. Gottschalk and G. A. Hedlund [1]). Many concepts and results of the theory of topological transformation groups can be found in the book by I. Yu. Bronshtein [5]. We shall present only several of them.

Let A be a collection of subsets of the group G – the so-called *admissible* subsets.

Definition 1.27.   A point $p \in X$ is a *recursive* point if, given any neighborhood U of the point p, there exists an admissible subset $A \in A$ such that

$$f(p, A) \subseteq U.$$

Definition 2.27.   A point $p \in X$ is a *locally recursive* point if, given any neighborhood U of the point p, there exist a neighborhood V of this point and an admissible subset $A \in A$ such that

$$f(V, A) \subseteq U.$$

Definition 3.27.   A point $p \in X$ is a *regionally recursive* point if, given any neighborhood U of the point p, there exists an admissible subset $A \in A$ such that, for arbitrary $g \in A$,

$$U \cap f(U, g) \neq \phi.$$

Definition 4.27.   Let X be a uniform space (see, e.g., N. Bourbaki [2]). A topological transformation group $[X, G, f]$ is *recursive* if, given any entourage α of the uniform structure of the space X, there exists an admissible subset $A \subseteq G$ such that $(p, q) \in \alpha$ for arbitrary points $p \in X, q \in f(p, A)$.

A subset $A \subseteq G$ is *syndetic* if there exists a compact $K \subseteq G$ such that $G = A + K$. It is not difficult to show that for $G = R$ the concept of a syndetic set coincides with the concept of a relatively dense set.

A subset $\Gamma \subseteq G$ is *saturated* if, given any compact $K \subseteq G$, there exist elements $g_1, g_2 \in G$ such that $g_1 + K + g_2 \subseteq \Gamma$. A subset $A \subseteq G$ is *extensive* if it intersects every saturated subgroup of the group G.

If $G = R$, then the set $A \subseteq R$ is extensive if and only if A contains a sequence which tends to $+ \infty$ and a sequence which tends to $- \infty$.

Definition 5.27.   If the expression "admissible set" in Definitions 1.27, 2.27, 3.27 and 4.27 is replaced by the expression "syndetic set" or "extensive set," then the expression "recursive point" is replaced by "almost periodic point" or "recurrent point" respectively.

This definition corresponds to the terminology frequently encountered in the American literature.

It is not difficult to note that in the case where X is a metric space and $G = R$ the concept of an almost periodic (recurrent, regionally recurrent) point in the sense of Definitions 1.27, 3.27 and 5.27 coincides with the concept of an almost recurrent (Poisson stable, nonwandering) point in the sense defined in the preceding chapters. Furthermore, if $X = f(p, G)$, then the concept of an almost periodic (recurrent) dynamical system in the sense of Definitions 4.27 and 5.27 coincides with the concept of an almost periodic (uniformly Poisson stable) motion in the sense of the definition in § 21. When reading journal articles the reader must, of course, take into consideration this variance in terminology.

We remark also that if X is a uniform space then we say that the transformation group $[X, G, f]$ is *equicontinuous* at the point $p \in X$ if the family of transformations $f(p, g_0)$ (for all $g_0 \in G$) of the space X into itself is equicontinuous at the point p, i.e., for any entourage α of the uniform structure of the space X there exists an index β such that for any point $q \in X$ such that $(p, q) \in \beta$ and any $g \in G$ we have

$$(f(p, g), f(q, g)) \in \alpha.$$

It is clear that this definition corresponds to the definition of Lyapunov stability (§ 22).

For topological transformation groups the following propositions, for instance, have been established.

THEOREM 1.27. *Let X be a regular space and let $p \in X$ be an almost periodic point. Then the closure of the orbit $f(p, G)$ is a minimal set.*

[See, e.g., P. S. Aleksandrov [2], page 303 for the definition of a regular space.]

THEOREM 2.27. *If X is a complete metric space and G is a connected abelian group, then all its points are regionally recurrent if and only if the set of all recurrent points is dense in X.*

THEOREM 3.27. *If X is a uniform space, the point p is almost periodic and equicontinuous, and the group G abelian, then all points in the closure of the orbit $f(p, G)$ are almost periodic.*

Recently, the concepts of *distalness* and *proximality* have attracted much interest among researchers. [See the articles by R. Ellis [1], R. Ellis and W. H. Gottschalk [1], and also the book by I. Yu. Bronshtein [5].]

A pair (p, q) $(p, q \in X)$ is *proximal* if, for any $\varepsilon > 0$, there exists an element $g \in G$ such that the inequality

$$\rho(f(p, g), f(q, g)) < \varepsilon$$

is satisfied.

A dynamical system is *distal* if for any two distinct points p and q in X there exists a number $\delta > 0$ such that, for all $g \in G$,

$$\rho(f(p, g), f(q, g)) > \delta.$$

§ 28 Discrete dynamical systems

Definition 1.28. A general dynamical system $[X, G, f]$ is *discrete* if the group G is cyclic, i.e., all its elements are multiples of an element $g_0 \in G$.

Example 1.28. Let $X = R^2$ be the plane and let $G = Z$ be the set of integers. We set into correspondence with the point $p \in R^2$ with the coordinates (x_0, y_0) and the integer k the point $q = f(p, k)$ with coordinates $(x_0 + k, y_0)$. In this case the orbit of the point p is a countable set of points distributed over the line $y = y_0$ and at integer-valued distances from the point p_0.

Note that to define the correspondence rule in the case of a cyclic group it obviously suffices to give $f(p, g_0)$, i.e., some homeomorphism of X onto X.

Example 2.28.   Let $X = R^2$ be the plane and let $G = Z$ be the set of integers. Here, 1 plays the role of g_0. Suppose the point p has polar coordinates (ρ_0, φ_0) and that $\rho_0 \neq 0$. We define the point $q = f(p, 1)$ so that its polar coordinates are $(\rho_0, \varphi_0 + \pi/2)$. In this case, $f(p, 4) = f(p, 0) = p$, the motion is periodic, and the orbit consists of four points. The origin of coordinates is considered a rest point.

Clearly, if one considers an ordinary dynamical system $[X, R, f]$, then the function $q = f(p, kt_0)$ ($t_0 \in R$ is a fixed number, $k = 0, \pm 1, \pm 2, \ldots$) defines a discrete dynamical system $[X, \{kt_0\}, f]$.

The set $S \subseteq X$ is a *global section* for the ordinary dynamical system $[X, R, f]$ if there exists a $t_0 \in R$ such that

$$S = f(P, \{kt_0\}),$$

where P is a set containing precisely one point from each of the trajectories $f(p, R) (p \in X)$.

Clearly, the ordinary dynamical system $[X, R, f]$ induces on its global section a discrete dynamical system

$$[S, \{kt_0\}, f].$$

It is interesting to ask under what conditions can a given discrete dynamical system be completed to a continuous dynamical system. N. P. Zhidkov [1] proved the following theorem.

THEOREM 1.28.   *If a discrete dynamical system is defined on a compactum* $K \subseteq R^n$, *then there exists on a subset of* R^{2n+1} *a continuous dynamical system which is defined by differential equations and has* K *as a global section.*

So-called *symbolic* dynamical systems are a special case of discrete dynamical systems. They are defined in the space S_m of two-sided sequences with m values ($m \geq 2$), e.g., the numbers $0, 1, \ldots, m-1$, in which case S_m consists of functions $p = p(i)$ which are defined on the set Z of all integers and whose values are the numbers $0, 1, \ldots, m-1$.

Let $p, q \in S_m$. Set $n(p, q) = 0$ if $p(0) \neq q(0)$,

$$n(p, q) = \sup\{n | p(i) = q(i) \text{ for } |i| < n\}$$

if $p(0) = q(0)$, and define a distance in S_m by the formula

$$\rho(p, q) = \frac{1}{1 + n(p, q)}.$$ (1.28)

Clearly, the sequence $\{p_n\}$ of points in S_n converges to the point $p_0 \in S_m$ if and only if, for any finite segment $[k_1, k_2] \subset Z$, there exists an index N such that for all $n > N$ and $i \in [k_1, k_2]$ the equalities

$$p_n(i) = p_0(i)$$

hold.

In this connection it is easy to see that the metric (1.28) is topologically equivalent to the Bebutov metric

$$\rho(p, q) = \sup_n \min \left\{ \max_{|i| < n} |p(i) - q(i)|; \frac{1}{n} \right\},$$

where n ranges over all positive integers.

Now define a symbolic dynamical system $[S_n, Z, f_0]$ by setting

$$f_0(p, 1) = q(p, q \in S_m),$$

where

$$q(i) = p(i + 1)$$

(a displacement of the sequence p by one unit to the left).

The following theorem, for instance, holds for a symbolic dynamical system.

THEOREM 2.28. *If the point $p \in S_m$ is either Poisson or Lyapunov stable, then it is periodic.*

It is suggested that the reader try to formulate the definitions of Poisson and Lyapunov stability and periodicity and prove this theorem.

Discrete and especially symbolic dynamical systems were considered by G. D. Birkhoff [3], G. D. Birkhoff and P. A. Smith [1], M. Morse [1], M. Morse and G. A. Hedlund [1, 2], N. P. Zhidkov [1], W. L. Reddy [1]. More general discrete dynamical systems defined by a single-valued (not necessarily a homeomorphic) mapping T of the topological space X into itself were studied by A. N. Sharkovsky [1–13]. The majority of the results obtained by him concern the case when X is a closed interval of the real line. The fundamentals of discrete systems in the case of a continuous mapping T are discussed in the article by K. Jacobs [1].

§ 29   Partially ordered dynamical systems

Definition 1.29.   A group G is (nontrivially) *partially ordered* if for any elements of G there is defined an order relation $g_1 > g_2$ (or, equivalently, $g_2 < g_1$) which possesses the following properties:

1.29. $g > g$ is impossible;
2.29. $g_1 > g_2$ and $g_2 > g_3$ imply $g_1 > g_3$;
3.29. If $g_1 > g_2$, then $g + g_1 > g + g_2$ and $g_1 + g > g_2 + g$ for any $g \in G$.
(See, e.g., A. G. Kurosh [1] or L. Fuks [1].)

An example of a partially ordered group is the group of complex numbers with respect to the operation of addition if one assumes that $a + bi > c + di$ for $a > c, b > d$ (group G_1) or for $a > c, b = d$ (group G_2).

Definition 2.29.   A general dynamical system $[X, G, f]$ whose group G is partially ordered is called a *partially ordered dynamical system*.

Example 1.29.   Let $X = R^2$ and let G be the group of vectors of a two-dimensional space or the complex numbers: G_1 or G_2. The correspondence rule can be defined as in Example 1.27.

The partial ordering of a group G allows one to rather easily carry over to these dynamical systems all of the fundamental concepts that were considered in the preceding chapters. For example, one can introduce the concepts of $\omega(\alpha)$-limit points, motions, Poisson stable motions, recurrent motions, almost periodic motions, and Lyapunov stable motions.

We now assume that the space X is metric and state some definitions and theorems.

Definition 3.29.   A point q is an *ω-limit point* for the motion $f(p, g)$ if for any $\varepsilon > 0$ and $g \in G$ there exists an element $g_1 > g$ such that

$$f(p, g_1) \in S(q, \varepsilon).$$

Definition 4.29.   A motion $f(p, g)$ is *almost recurrent* if for any $\varepsilon > 0$ there exists an element $g_\varepsilon > 0$ ($g_\varepsilon \in G$) such that, for arbitrary $g_0 \in G$,

$$p \in S(f(p, [g_0, g_0 + g_\varepsilon]), \varepsilon),$$

where

$$[g_0, g_0 + g_\varepsilon] = \{g \in G | g_0 \leq g \leq g_0 + g_\varepsilon\}.$$

Definition 5.29. A motion $f(p, g)$ is *recurrent* if for any $\varepsilon > 0$ there exists an element $g_\varepsilon > 0$ $(g_\varepsilon \in G)$ such that, for arbitrary $g_0 \in G$,

$$f(p, G) \subseteq S(f(p, [g_0, g_0 + g_\varepsilon]), \varepsilon).$$

Example 2.29. As the space X consider the square $0 \leq x \leq 1, 0 \leq y \leq 1$ in the xy-plane. Let φ be a homeomorphism of the segment $[0, 1]$ onto the extended real line $[-\infty, +\infty]$ (the real line to which have been adjoined two points denoted by $-\infty$ and $+\infty$) and let $G = G_1$ or $G = G_2$, the group of complex numbers $\{a + bi\}$. Set

$$f((x, y), a + bi) = (\varphi^{-1}(\varphi(x) + a), \varphi^{-1}(\varphi(y) + b)).$$

Clearly, the points $(0, 0), (0, 1), (1, 0)$ and $(1, 1)$ are rest points. The orbit of any interior point of the square X is the entire interior of the square; and the orbits of points belonging to precisely one side of the square X coincide with the side corresponding to it (without vertices).

It is easy to show that if p is an interior point of the square X, then $\Omega_p = \{(1, 1)\}$ when $G = G_1$ and $\Omega_p = \phi$ when $G = G_2$.

Definition 6.29. A nontrivial partially ordered group G is *directed* if for any pair of elements g_1, g_2 of G there exists an element $g \in G$ such that g is greater than each of them: $g > g_1, g > g_2$.

Clearly, a group G_1 is directed whereas G_2 does not possess this property. This also explains the fact that in Example 2.29 the set $\Omega_p = \phi$ when $G = G_2$.

THEOREM 1.29. *If the group G is directed, then the ω-limit set of a positively Lagrange stable motion is nonempty.*

THEOREM 2.29. *For a directed group G, all motions in a compact minimal set are recurrent.*

Definition 7.29. The *integral continuity condition* is said to be satisfied in a partially ordered dynamical system $[X, G, f]$, if, for any element

$g_0 > 0$, number $\varepsilon > 0$ and point $p \in X$, there exists a $\delta > 0$ such that

$$\rho(f(q, g), f(p, g)) < \varepsilon,$$

for arbitrary $q \in S(p, \delta)$ and $g \in [0, g_0]$.

Note that in contrast to ordinary dynamical systems $[X, R, f]$ the integral continuity condition may not be satisfied for partially ordered dynamical systems.

THEOREM 3.29. *If the integral continuity condition is satisfied, then the closure of the trajectory of an almost recurrent motion is a minimal set.*

Partially ordered dynamical systems were first introduced by E. A. Barbashin [2, 3, 9]. They were studied in detail by A. M. Stakhi [1-5].

§ 30   Dispersive dynamical systems

Let X be a metric space and let $A \subseteq X, B \subseteq X$ be two closed bounded subsets of X. The number

$$\beta(A, B) \equiv \sup_{a \in A} \rho(a, B)$$

is called the *semi-deviation* of the set A from the set B. The number

$$\alpha(A, B) \equiv \max\,(\beta(A, B), \beta(B, A))$$

is called the *deviation* between the sets A and B.

Let X be a given metric space and suppose that to each point p of this space and moment of time $t \in R$ there is set into correspondence a non-empty closed compact subset $f(p, t) \subseteq X$ whereby, if we let

$$f(A, K) \equiv \bigcup_{p \in A, t \in K} f(p, t),$$

then

    1.30. $f(p, 0) = p$ for every point $p \in X$.
    2.30. If $q \in f(p, t)$, then $p \in f(q, -t)$.
    3.30. If $t_1 t_2 > 0$, then $f(f(p, t_1), t_2) = f(p, t_1 + t_2)$.

4.30. $\lim_{\substack{p \to p_0 \\ t \to t_0}} \beta(f(p, t), f(p_0, t_0)) = 0.$

If this is the situation, then we say that a *dispersive* dynamical system has been defined.

The sets $f(p, R)$, $f(p, R^+)$ and $f(p, R^-)$ are called the *funnel*, *positive semifunnel* and the *negative semifunnel*, respectively.

Example 1.30.   Consider an arbitrary point $p(x_0, y_0)$ in the plane R^2 and any moment of time $t \in R$. We define the set $f(p, t)$ as follows (see Fig. 16):

$$f(p, t) = \bigcup_{|y - y_0| \le t} q(x_0 + t, y).$$

It is easy to verify that in this case all the conditions for a dispersive dynamical system are satisfied.

Definition 1.30.   A point $p_0 \in X$ is a *point of continuity* if, for any $t \in R$,

$$\lim_{p \to p_0} \alpha(f(p, t), f(p_0, t)) = 0.$$

Definition 2.30.   The *integral continuity condition* is satisfied if, for any $p \in X$, $\varepsilon > 0$ and $T > 0$, there exists a $\delta > 0$ such that, for all $q \in S(p, \delta)$ and $|t| \le T$,

$$\alpha(f(q, t), f(p, t)) < \varepsilon.$$

The concepts of a point of continuity and of the integral continuity condition, which as in the case of a partially ordered dynamical system does not follow from the corresponding axioms of a dynamical system, play an important role in the theory of dispersive dynamical systems.

Dispersive dynamical systems possess the following special properties:

1. Concepts which were two-sided (e.g., invariance, minimality) for ordinary dynamical systems are not considered here. To clarify this fact it suffices to note that $q \in f(p, R)$ does not imply the inclusion $f(q, R) \subseteq f(p, R)$ (see Fig. 16).

2. Many concepts (invariance, dynamical limit points, Poisson stability, and others) admit a great variance in their definitions. We can give, e.g., the following definitions.

Definition 3.30.   A set $A \subseteq X$ is *positively semi-invariant* (*positively quasi-invariant*) if, for any $t \in R^+$,

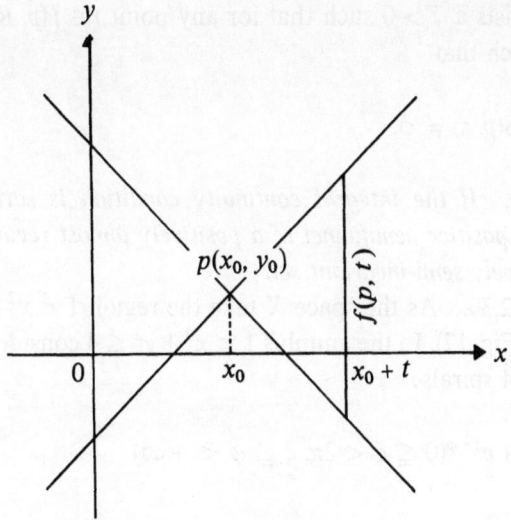

Figure 16

$f(A, t) \subseteq A \,[f(A, t) \supseteq A]$.

A set $A \subseteq X$ is *positively pseudo-invariant* if, for any $p \in A$ and $t \in R^+$,

$f(p, t) \cap A \neq \phi$.

Definition 4.30.   A nonempty closed positively semi-invariant (quasi-invariant, pseudo-invariant) set is a *minimal* positively semi-invariant (quasi-invariant, pseudo-invariant) set if it does not contain a nonempty closed positively semi-invariant (quasi-invariant, pseudo-invariant) proper subset.

THEOREM 1.30.   *If the set $A \subseteq X$ is positively semi-invariant and $\bar{A} \setminus A$ consists of points of continuity, then $\bar{A}$ is positively semi-invariant.*

THEOREM 2.30.   *A necessary and sufficient condition for a closed positively semi-invariant set Σ, consisting of points of continuity, to be a minimal positively semi-invariant set is that, for any point $p \in \Sigma$, the closure of its positive funnel coincide with Σ.*

Definition 5.30.   A point p is *positively almost recurrent* if for any

141

$\varepsilon > 0$ there exists a $T > 0$ such that for any point $r \in f(p, R^+)$ there exists a $t \in [0, T]$ such that

$$f(r, t) \cap S(p, \varepsilon) \neq \phi.$$

THEOREM 3.30.   *If the integral continuity condition is satisfied, then the closure of the positive semifunnel of a positively almost recurrent point is a minimal positively semi-invariant set.*

   Example 2.30.   As the space X take the region $1 < x^2 + y^2 \leq 9$ in the xy-plane (see Fig. 17). In the annulus $1 < x^2 + y^2 \leq 4$ consider the following two families of spirals:

$$l_\xi : r = 1 + e^{\xi - \varphi}(0 \leq \xi < 2\pi, \xi \leq \varphi < +\infty)$$

and

$$L_\eta : r = 1 + e^{\varphi - \eta}(-2\pi < \eta \leq 0, -\infty < \varphi \leq \eta),$$

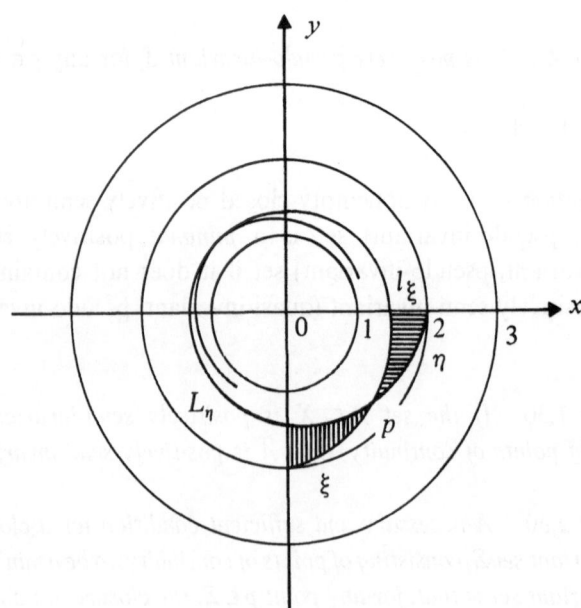

Figure 17

where φ is the polar angle. Consider an arbitrary point $p(r_0, \varphi_0)$ of the annulus $1 < x^2 + y^2 \leqq 4$. One of the spirals l_ξ and one of the spirals L_η pass through it. Denote by $f(p, t)$ the segment of the radius vector $\varphi = \varphi_0 + t$ included between these two spirals if

$$|t| \leqq \ln (r_0 - 1)$$

and between the spiral $l_\xi(L_\eta)$ and the circle $x^2 + y^2 = 4$ if

$$t > \ln (r_0 - 1)(t < -\ln (r_0 - 1)).$$

An analogous construction is made in the annulus $4 < x^2 + y^2 \leqq 9$. In this example the point p_0 with the coordinates $r_0 = 2$, $\varphi_0 = 0$ is positively almost recurrent, however the closure of the semifunnel $f(p_0, R^+)$ is not a minimal positively semi-invariant set since it is not even positively semi-invariant. Here the situation is that the integral continuity condition is not satisfied since the points of the circle $x^2 + y^2 = 4$ are not even points of continuity.

An important concept in the theory of dispersive dynamical systems is that of a motion.

Definition 6.30.   A single-valued continuous mapping $q = \varphi_p(t)$ of the real line into the space X is a *motion* emanating from the point p if $\varphi_p(0) = p$, and for any $t_1 < t_2 (t_1, t_2 \in R)$ the inclusion

$$\varphi_p(t_2) \in f(\varphi_p(t_1), t_2 - t_1)$$

holds.

The set $\varphi_p(R) \equiv \bigcup_{t \in I} \varphi_p(t)$ is called the *trajectory* of this motion.

The theory of dispersive dynamical systems was worked out by E. A. Barbashin [1, 5, 6, 8] and is a generalization of the theory of systems of differential equations without uniqueness assumptions. The papers by M. I. Minkevich [1–3], B. M. Budak [1, 2] and M. S. Izman [3, 5] are also devoted to the investigation of dispersive dynamical systems.

As E. A. Barbashin and Yu. I. Alimov [1, 2] showed, dispersive dynamical systems arise also in considering systems of differential equations with multi-valued right members (equations with contingencies).

When considering nonautonomous systems of differential equations, V. I. Zubov [1] and E. Roxin [1, 2] noted that under definite restrictions the solutions of these systems also define dispersive dynamical systems.

Nonautonomous systems from the viewpoint of the theory of dynamical systems were investigated in the works of V. A. Belov [1], O. O. Hájek [1], L. G. Deysach and G. R. Sell [1], L. Markus [1], R. K. Miller [1], V. M. Millionshchikov [1, 2], Z. Opial [1, 2], G. R. Sell [1, 2], and B. A. Shcherbakov [5, 6, 8].

I. Yu. Bronshtein [1–4] developed the theory of general dispersive dynamical systems, proposing in this connection the concept of a dynamical system without uniqueness as semigroups of multiple -valued mappings of a topological space. The transition to a semigroup is natural since in the theory of dispersive dynamical systems one actually considers the properties of semigroups of mappings $f(p, t)$ for $t \geq 0$ and parallel with that the properties of semigroups of mappings $f(p, t)$ for $t \leq 0$.

N. K. Cheban [1–3] considered partially ordered dispersive dynamical systems as partially ordered semigroups of multi-valued mappings of the space.

LITERATURE

[An asterisk * indicates an addition for the American edition.]

ALEKSANDROV, P. S.
 [1] Combinatorial topology, Moscow-Leningrad, Gostekhizdat, 1947 (Russian); Rochester, Graylock Press, 1960, 1961, 1962.
 [2] Introduction to the general theory of sets and functions, Moscow-Leningrad, GITTL, 1948 (Russian); New York, Chelsea, in preparation.

ALMUKHAMEDOV, M. I.
 [1] *The space of semi-periodic functions in the theory of dynamical systems*, Uch. Zap. Kazansk. Ped. In-ta, 10 (1955) 29–56 (Russian); Ref Zh M, 1956, 5237.

AUSLANDER, J. and P. SEIBERT
 [1] *Prolongations and generalized Liapunov functions*, International Symposium on Nonlinear Differential Equations and Nonlinear Mechanics, N.Y.-London, Academic Press (1963) 454–462; MR 27 # 416.
 [2] *Prolongations and stability in dynamical systems*, Ann. Inst. Fourier (Grenoble) 14 (1964) 237–267; MR 31 # 455.

BARBASHIN, E. A.
 [1] *Sur certaines singularités qui surviennent dans un système dynamique quand l'unicité est en défaut*, Doklady AN SSSR, 41 (1943) 139–147 (Russian summary); MR 6 # 86.
 [2] *Sur la conduite des points sous les transformations homéomorphes de l'espace*, Doklady AN SSSR 51 (1946) 3–5; MR 8 # 34.
 [3] *Sur la classification des multiplicités intégrales d'un système d'équations en différentielles totales*, Doklady AN SSSR 55 (1947) 279–286 (Russian summary); MR 8 # 589.
 [4] *On homomorphisms of dynamical systems*, Doklady AN SSSR 61 (1948) 429–432 (Russian).
 [5] *On the theory of generalized dynamical systems*, Uch. Zap. Moskovsk. Un-ta, No. 135 (Matematika) 2 (1948) 110–133 (Russian).
 [6] *On the theory of systems of non-single-valued mappings of a topological space*, Uch. Zap. Uralsk. Un-ta 7 (1950) 54–60 (Russian).
 [7] *On homomorphisms of dynamical systems*, Matem. Sbornik 27 (1950) 455–470 and 29 (1951) 501–518 (Russian); MR 12 # 422 and MR 13 # 473.
 [8] *Dispersive dynamical systems*, Uspekhi Matem. Nauk 5 (1950) 138–139 (Russian); MR 12 # 336.
 [9] *On the behavior of points under homeomorphic transformations of a space. (Generalization of Birkhoff's theorems)*. Trudy Uralsk. Politekhn. In-ta 51 (1954) 4–11 (Russian); MR 17 # 1230.

145

Literature

BARBASHIN, E. A. and YU. M. ALIMOV
  [1] *On the theory of dynamical systems with multi-valued and discontinuous characteristics*, Doklady AN SSSR, **140** (1961) 9–11 (Russian); MR **25** # 4209.
  [2] *On the theory of relay differential equations*, Izv. VUZov (Matematika), **1** (1962) 3–13 (Russian); MR **25** # 2272.

BEBUTOV, M. V.
  [1] *Sur les systèmes dynamiques dans l'espace des fonctions continues*, Doklady AN SSSR **27** (1940) 904–906; MR **2** # 225.
  [2] *On dynamical systems in the space of continuous functions*, Byulletin Moskovsk. Un-ta (Matematika) **2** (1941) 1–52 (Russian).

BELOV, V. A.
  [1] *Asymptotic trajectories of non-autonomous dynamical systems*, Dif. Uravneniya **3** (1967) 226–231 (Russian); MR **35** # 3184.

BENDIXSON, I.
  [1] *Sur les courbes définies par des équations différentielles*, Acta Math. **24** (1901) 1–88. There is a Russian translation of Chapter I: *On curves defined by differential equations*, Uspekhi Matem. Nauk **9** (1941) 191–211.

BHATIA, N. P.
  [1] *Stability and Liapunov functions in dynamical systems*, Contributions to Diff. Equations, New York, Wiley **3** (1964) 175–188; MR **29** # 319.

BHATIA, N. P., A. C. LAZER, and G. P. SZEGÖ
  [1] *On global weak attractors in dynamical systems*, J. Math. Analysis and Appl. **16** (1966) 544–552; MR **34** # 5076; Ref Zh M, 1967, 10 Б324.
  [2] *On global weak attractors in dynamical systems*, Diff. Equations and Dynamical Systems, New York-London, Acad. Press, 1967, 375–379.

BHATIA, N. P. and G. P. SZEGÖ
  [1] Dynamical systems: stability theory and applications, Berlin-Heidelberg-New York, Springer-Verlag, 1967; MR **36** # 2917; Ref Zh M, 1968, 6 Б 252 K.
  *[2] *Stability Theory of Dynamical Systems*, Grundlehren der math. Wiss., Berlin, Heidelberg, N.Y., Springer Verlag, **161**, 1970; MR **44** # 7077.

BIRKHOFF, G. D.
  [1] Dynamical systems, Moscow, GITTL (1941) (Russian); New York, AMS Colloquium Publications **9** (1927), first edition.
  [2] *Quelques théorèmes sur les mouvements des systèmes dynamiques*, Bull. Soc. Math. de France **40** (1912) 305–323.
  [3] Collected Mathematical Papers, Vol. 1–3, New York, AMS, 1950; New York, Dover Publications, 1968.

BIRKHOFF, G. D. and P. A. SMITH
  [1] *Structure analysis of surface transformations*, J. Math. Pures et Appl. **9** (1928) 345–379.

146

BOHR, H.

[1] Almost periodic functions, Moscow, GITTL (1934) (Russian); New York, Chelsea, 1947 (translated by Harvey Cohn).

BOHR, H. and W. FENCHEL

[1] *Ein Satz über stabile Bewegungen in der Ebene*, Harold Bohr, Collected Math. Works, Copenhagen, **2** (1952) 38.

BOURBAKI, N.

[1] General topology. Numbers and groups and spaces connected with them, Moscow, GIFML (1959) (Russian); Ref Zh M, 1961, 7 A 401 K.

[2] General topology. Fundamental structures, Moscow, GIFML (1958) (Russian); Ref Zh M, 1961, 7 A 400 K.

BRONSHTEIN, I. YU.

[1] *On dynamical systems without uniqueness as semigroups of non-single-valued mappings of a topological space*, Doklady AN SSSR **144** (1962) 954–957 (Russian); Ref Zh M, 1963, 5 A 328. Soviet Math. Doklady, **3** (1962) 824–827 (English translation).

[2] *Recurrence, periodicity and transitivity in dynamical systems without uniqueness*, Doklady AN SSSR **151** (1963) 15–18 (Russian); MR **30** # 573, Ref Zh M, 1963, 11 Б 527. Soviet Math. Doklady, **4** (1963), 889–892 (English translation).

[3] *On dynamical systems without uniqueness as semigroups of non-single-valued mappings of a topological space*, Izv. AN MSSR **1** (1963) 3–18 (Russian); Ref Zh M, 1964, 3 Б 266.

[4] *Recurrent points and minimal sets in dynamical systems without uniqueness*, Izv. AN MSSR **7** (1965) 14–21 (Russian); Ref Zh M, 1966, 6 Б 282.

[5] *Minimal groups of transformations*, RIO AN MSSR (1969) (Russian); Ref Zh M, 1970, 2 A 446 K.

*[6] *On solutions of algebraic equations which are recurrent in the sense of Birkhoff*, Matem. Zametki **13** (1973) 617–623 (Russian); Ref Zh M, 1973, 8 Б 744.

BRONSHTEIN, I. YU. and B. A. SHCHERBAKOV

[1] *Certain properties of Lagrange stable funnels of generalized dynamical systems*, Izv. AN MSSR **5** (1962) 99–102 (Russian); Ref Zh M, 1963, 3 Б 220.

BUDAK, B. M.

[1] *Dispersive dynamical systems*, Vestnik Moskovsk. Un-ta **8** (1947) 135–137 (Russian); MR **10** # 309.

[2] *The concept of motion in a generalized dynamical system*, Uch. Zap. Moskovsk. Un-ta, No. 155 (Matematika) **5** (1952) 174–194 (Russian); MR **18** # 407.

CHEBAN, N. K.

[1] *Recurrent points and minimal sets in partially ordered dispersive dynamical systems*, Matem. Issled., RIO AN MSSR **2** (1967) 147–159 (Russian); Ref Zh M, 1968, 4 Б 731.

[2] *Uniform approximation of sets in partially ordered dispersive dynamical systems*, Izv. AN MSSR **8** (1967) 87–91 (Russian); Ref Zh M, 1968, 7 Б 309.

[3] *Stability in partially ordered dispersive dynamical systems*, Matem. Issled., RIO AN MSSR **1** (1968) 149–158 (Russian); Ref Zh M, 1969, 1 Б 331.

Literature

*[4] *Invariant sets in partially ordered semigroups of multi-valued mappings of a space,*
   Matem. Issled., RIO AN MSSR 3 (1970) 172–181 (Russian); Ref Zh M, 1971, 6 A 515.

CHELYSHEVA, L. A.
   [1] *Topological classification of closed invariant sets,* Matem. Issled. RIO AN MSSR 1
   (1968) 184–197 (Russian); Ref Zh M, 1969, 1 Б 332.

CHURCHILL, R. C.
   *[1] *Isolated invariant sets in compact metric spaces,* J. of Diff. Equations 12 (1972) 330–352;
   Ref Zh M, 1973, 5 Б 284.

CONLEY, C. C.
   *[1] *On the ultimate behavior of orbits with respect to an unstable critical point. I. Oscil-
   lating, asymptotic, and capture orbits,* J. of Differential Equations 5 (1969) 136–158;
   MR 40 # 4532.
   *[2] *On the continuation of invariant sets of a flow,* Actes du congrès int. des math., Gauthier-
   Villars, Paris 2 (1971) 909–913; Ref Zh M, 1972, 3 Б 237.

CONLEY, C. C. and R. W. EASTON
   *[1] *Isolated invariant sets and isolating blocks,* Trans. Amer. Math. Soc. **158** (1971)
   35–61; MR **43** # 5551.

DEYSACH, L. G. and G. R. SELL
   [1] *On the existence of almost periodic motions,* Michigan Math. J. **12** (1965) 87–95; MR
   30 # 3279.

ELLIS, R.
   [1] *Distal transformation groups,* Pacific J. Math. **8** (1958) 401–405; MR **21** # 96.
   *[2] *Lectures on topological dynamics,* W. A. Benjamin, New York, 1969; MR **42** # 2463.

ELLIS, R. and W. H. GOTTSCHALK
   [1] *Homomorphisms of transformation groups,* Trans. AMS **94** (1960) 258–271; MR **23**
   # A 960.

ENGLAND, J. W.
   [1] *A characterization of orbits,* Proc. AMS **17** (1966) 207–209; MR **32** # 3052.

FILIPPOV, A. F.
   [1] *Differential equations with discontinuous right member,* Matem. Sbornik **51** (1960)
   99–128 (Russian); MR **22** # 4846.

FOLAND, N. E.
   [1] *The structure of the orbits and their limit sets in continuous flows,* Pacific J. Math. **13**
   (1963) 563–570; MR **28** # 301.

FRANKLIN, P.
   [1] *Almost periodic recurrent motions,* Math. Zeitschrift **30** (1929) 325–331.

FUKS, L. (= L. FUCHS)
   [1] Partially ordered algebraic systems, Moscow, Mir (1965) (Russian); Ref Zh M, 1965,
   10 A 268 K; Pergamon Press, Oxford and Addison-Wesley, Reading, 1963; MR **30**
   # 2090.

GOTTSCHALK, W. H.
[1] Bibliography for topological dynamics (4th ed.), Middletown, Wesleyan Univ., 1969; MR **41** # 2655.

GOTTSCHALK, W. H. and G. A. HEDLUND
[1] Topological Dynamics, Providence, AMS Colloq. Publ. **36** (1955); MR **17** # 650.

HAHN, W.
*[1] Stability of Motion, New York, Springer-Verlag, 1967; MR **36** # 6716.

HÁJEK, O.
[1] *Flows and periodic motions*, Comment. Math. Univ. Carolinae **6** (1965) 164–178; MR **33** # 3284.

HALE, J. and J. P. LASALLE (editors)
*[1] Differential Equations and Dynamical Systems, New York, Academic Press, 1967; MR **36** # 3; Ref Zh M, 1968, 7 Б 142 K.

HILMY, H. F.
[1] *Sur les centres d'attraction minimaux dans les systèmes dynamiques*, Comp. Math. **3** (1936) 227–238.

IZMAN, M. S.
[1] *Strong instability and attractor sets in dynamical systems*, Izv. AN SSSR **8** (1967) 81–86 (Russian); Ref Zh M, 1968, 7 Б 308.
[2] *On stable and attracting sets in dynamical systems*, Matem. Issled., RIO AN MSSR **3** (1968) 159–166 (Russian); RefZh M, 1969, 7 Б 253.
[3] *Stability and attractor sets in dispersive dynamical systems*, Matem. Issled., RIO AN MSSR **3** (1968) 60–78 (Russian); Ref Zh M, 1969, 7 Б 254.
[4] *Stability of sets and attractors in dispersive dynamical systems*, Matem. Issled., RIO AN MSSR **4** (1968) 51–77 (Russian); Ref Zh M, 1969, 11 Б 354.
[5] *Application of Lyapunov's second method to investigate the stability and asymptotic stability of sets in dispersive dynamical systems*, Dif. Uravneniya **5** (1969) 1207–1217 (Russian); Ref Zh M, 1969, 12 Б 361.
*[6] *On the asymptotic stability of sets in dispersive dynamical systems*, Dif. Uravneniya **7** (1971) 615–621 (Russian); Ref Zh M, 1971, 10 Б 272.

JACOBS, K.
[1] *Einige Grundbegriffe der topologischen Dynamik*, Math.-phys. Semesterber. **14** (1967) 129–150; Selecta Mathematica, IV, Heidelberger Taschenbücher, Springer-Verlag, Berlin-Heidelberg, New York, 1–30; Ref Zh M, 1968, 5 A 477.

KAKUTANI, S.
[1] *A proof of Bebutov's theorem*, J. of Diff. Equations **4** (1968) 194–201; MR **37** # 1734.

KRASOVSKY, N. N.
[1] Some problems in the theory of the stability of motion, Moscow, GIFML (1959) (Russian); MR **21** # 5047.
*[2] Stability of Motion, Palo Alto, Stanford Univ. Press, 1963 (English translation of [1]).

149

Literature

KUROSH, A. G.
[1] Lectures on general algebra, Moscow, GIFML (1962) (Russian) MR **25** # 5097; New York, Chelsea (1965); Oxford-Edinburgh-New York, Pergamon (1965); MR **31** # 3483.

LADIS, N. N.
[1] *The absence of bilateral attraction points*, Dif. Uravneniya **4** (1968) 1157 (Russian); MR **37** # 4360.
[2] *On sets of bilateral attraction*, Dif. Uravneniya **6** (1971) 365–367 (Russian); MR **44** # 4730.

LYAPUNOV, A. M.
[1] General problem of the stability of motion, Moscow-Leningrad, GITTL (1950) (Russian). First edition, Kharkov (1892) (Russian).

LYUSTERNIK, L. A. and V. I. SOBOLEV
[1] Elements of functional analysis, Moscow-Leningrad, GITTL (1951) (Russian); New York, Ungar (1955).

MARKOV, A. A.
[1] *Sur une propriété générale des ensembles minimaux de M. Birkhoff*, Comptes r. Acad sci., Paris **193** (1931) 823–825.
[2] *Stabilität im Liapounoffschen Sinne und Fastperiodizität*, Mat. Zeitschr. **36** (1933) 708–738.

MARKUS, L.
[1] *Asymptotically autonomous differential systems*, Ann. Math. Studies, Princeton U. Press, Princeton **36** (1956) 17–29; MR **18** # 394.

MILLER, R. K.
[1] *Almost periodic differential equations as dynamical systems with applications to the existence of a. p. solutions*, J. of Differential Equations **1** (1965) 337–345; MR **32** # 262.

MILLER, R. K. and G. R. SELL
*[1] *Volterra integral equations and topological dynamics*, Memoirs of the Amer. Math. Soc. **102** (1970); MR **44** # 5579.

MILLIONSHCHIKOV, V. M.
[1] *Recurrent and almost periodic limit trajectories of nonautonomous systems of differential equations*, Doklady AN SSSR **161** (1965) 43–44 (Russian); MR **30** # 4043.
[2] *On recurrent and almost periodic limit trajectories of nonautonomous systems*, Dif. Uravneniya **4** (1968) 1555–1559 (Russian); Ref Zh M, 1969, 2 Б 228.

MINKEVICH, M. I.
[1] *The theory of integral funnels in generalized dynamical systems without a hypothesis of uniqueness*, Doklady AN SSSR **59** (1948) 1049–1052 (Russian); MR **9** # 449.
[2] *The theory of integral funnels in dynamical systems without uniqueness*, Uch. Zap. Moskovsk. Un-ta, No. 135 (Matematika) **2** (1948) 134–151 (Russian); MR **11** # 443.

[3] *Closed integral funnels in generalized dynamical systems without a hypothesis of uniqueness*, Uch. Zap. Moskovsk. Un-ta (Matematika) **163** (1952) 73–88 (Russian); Doklady AN SSSR **60** (1948) 341–343 (Russian); MR **9** # 517. Review of German translation: MR **17** # 363.

MONTGOMERY, J. T.
[1] Cohomology of isolated invariant sets under perturbation, J. of Diff. Equations **13** (1973) 257–299.

MORSE, M.
[1] Symbolic dynamics, Lectures by Marston Morse, 1937–1938. Princeton, Inst. for Advanced Study, 1966.

MORSE, M. and G. A. HEDLUND
[1] *Symbolic dynamics. I.* Amer. J. of Math. **60** (1938) 815–866.
[2] *Symbolic dynamics. II. Sturmian trajectories*, Amer. J. of Math. **62** (1940) 1–42; MR **1** # 123.

NEMYTSKY, V. V.
[1] *Systèmes dynamiques sur une multiplicité intégrale limité*, Doklady AN SSSR **47** (1945) 535–558 (Russian summary); MR **7** # 255.
[2] *Les systèmes dynamiques généraux*, Doklady AN SSSR **53** (1946) 491–498 (Russian summary); MR **8** # 280.
[3] *On the theory of orbits of general dynamical systems*, Matem. Sbornik **23** (1948) 161–186 (Russian); MR **10** # 259.
[4] *Topological problems in the theory of dynamical systems*, Uspekhi Matem. Nauk **4** (1949) 91–153 (Russian); MR **11** # 526. English translation, AMS Translation No. 103 (1954) 1–85.
[5] *Generalizations of the theory of dynamical systems*, Uspekhi Matem. Nauk **5** (1950) 47–59 (Russian); MR **12** # 34.
[6] *Seminar on the qualitative theory of differential equations at MGU*, Uspekhi Matem. Nauk **12** (1957) 235–239 (Russian); Ref Zh M, 1958, 4371.
[7] *Topological classification of singular points and generalized Lyapunov functions*, Dif. Uravneniya **3** (1967) 359–370 (Russian); Ref Zh M, 1967, 12 Б 265.

NEMYTSKY, V. V. and V. V. STEPANOV
[1] Qualitative theory of differential equations, 1st ed., Moscow-Leningrad, GITTL, 1947; 2nd ed., Moscow-Leningrad, GITTL, 1949 (Russian); MR **10** # 612; Princeton, Princeton Univ. Press, 1960; MR **22** # 12258.

OPIAL, Z.
[1] *Sur les solutions presque-périodiques d'une classe d'équations différentielles*, Ann. Polon. Math. **9** (1960) 157–181; MR **23** # A 390.
[2] *Sur une équation différentielle presque-périodique sans solution presque-périodique*, Bull. Acad. Polon. Sci. (sér. sci. math., astron. et phys.) **9** (1961) 673–676; MR **24** # A 2707.

Literature

POINCARÉ, H.

[1] *On curves which are defined by differential equations*, Moscow-Leningrad, GITTL, 1947 (with appendices) (Russian). *Sur les courbes définies par les équations différentielles*, Journ. de Math. Pures Appl. 1 (1885) 167–244.

[2] Les méthodes nouvelles de la mécanique céleste, 1 (1892), 2 (1893), 3 (1899), Paris, Gauthier-Villars; reprinted by Dover Publications, Inc., New York, 1957; Ref Zh M, 1961, 4 Б 356 K.

PONTRYAGIN, L. S.

[1] Continuous groups, 1st ed., Moscow, ONTI, 1938; 2nd ed., Moscow, GITTL, 1954 (Russian); Topological groups, Princeton, Princeton Univ. Press, 1939; Leipzig, B. G. Teubner, 1957, 1958; Ref Zh M, 1956, 2004 K.

PRIVALOV, I. I.

[1] Introduction to the theory of functions of a complex variable, Moscow, GITTL, 1954 (Russian); Ref Zh M, 1955, 3178 K.

REDDY, W. L.

[1] *Lifting expansive homeomorphisms to symbolic flows*, Math. Systems Theory, 2 (1968) 91–92; MR **36** # 7127.

ROXIN, E. O.

[1] *On generalized dynamical systems defined by contingent equations*, J. of Diff. Equations 1 (1965) 188–205; MR **34** # 1638.

[2] *Local definition of generalized control systems*, Michigan Math J. **13** (1966) 91–96; MR **32** # 9078.

ROZHKO, V. F.

*[1] *On the theory of discontinuous dynamical systems. I. Invariant and dynamical limit sets. Poisson stability*, Matem. Issled., RIO AN MSSR 3 (1969) 63–73 (Russian); MR **41** # 7665.

*[2] *On the theory of discontinuous dynamical systems. II. Minimal sets, recurrent and almost recurrent motions*, Matem. Issled., RIO AN MSSR 1 (1970) 102–117 (Russian); MR **43** # 2693.

*[3] *On the theory of discontinuous dynamical systems. III. Almost periodic motions*, Matem. Issled., RIO AN MSSR 2 (1970) 157–167 (Russian); MR **43** # 2694.

*[4] *On a class of almost periodic motions in systems with thrusts*, Dif. Uravneniya **8** (1972) 2012–2022 (Russian); Ref Zh M, 1973, 3 Б 334.

*[5] *On almost recurrent and recurrent motions of discontinuous dynamical systems*, Dif. Uravneniya **9** (1973) 1826–1830 (Russian); Ref Zh M, 1974, 2 Б 297.

SAITO, T.

*[1] *Isolated minimal sets*, Funkcial. Ekvac. **11** (1968) 155–167; MR **40** # 1669.

*[2] *On a compact invariant set isolated from minimal sets*, Funkcial. Ekvac. **12** (1969) 193–203; MR **41** # 2657.

*[3] *A supplement to the paper "On a compact invariant set isolated from minimal sets"*, Funkcial. Ekvac. **13** (1970) 127–129; MR **43** # 5512.

*[4] *On the structure of compact dynamical systems*, Funkcial. Ekvac. **13** (1971) 147–170; MR **43** # 7733.

SEIBERT, P.

*[1] *Stability in dynamical systems,* In the collection *Stability problems of solutions of differential equations,* Oderisi, Gubbio, 1966, 73–94.

SEIBERT, P. and P. TULLEY

[1] *On dynamical systems in the plane,* Arch. Math. **18** (1967) 290–292; MR **36** # 873; Ref Zh M, 1968, 4 A 421.

SELL, G. R.

[1] *Nonautonomous differential equations and topological dynamics, I. The basic theory,* Trans. AMS **127** (1967) 241–262; MR **35** # 3187a; Ref Zh M, 1968, 8 Б 331.

[2] *Nonautonomous differential equations and topological dynamics, II. Limiting equations,* Trans. AMS **127** (1967) 263–283; MR **35** # 3187b; Ref Zh M, 1968, 8 Б 332.

*[3] *Topological dynamics and ordinary differential equations,* Van Nostrand Reinhold Math. Studies # 33, London, 1971; Ref Zh M, 1972, 4 Б 434 K, 1973, 11 Б 237 K.

*[4] *Topological dynamical techniques for differential and integral equations,* In the collection *Ordinary differential equations,* 1971, NRL-MRC Conference, Acad. Press, New York, 1972, 287–304.

SHARKOVSKY, O. M. (A. N.)

[1] *Non-wandering points and the center of a continuous transformation of a line into itself,* Dop. AN URSR 7 (1964) 865–868 (Ukrainian); Ref Zh M, 1964, 12 Б 221.

[2] *Coexistence of cycles of a continuous transformation of a line into itself,* Ukr. Matem. Zh. **16** (1964) 61–71 (Russian); Ref Zh M, 1964, 10 Б 48.

[3] *On attracting and attracted sets,* Doklady AN SSSR **160** (1965) 1036–1038 (Russian); MR **32** # 6419.

[4] *On cycles and the structure of a continuous transformation,* Ukr. Matem. Zh. **17** (1965) 104–111 (Russian); MR **32** # 4213.

[5] *On a classification of fixed points,* Ukr. Matem. Zh. **17** (1965) 80–95 (Russian); MR **33** # 4915.

[6] *On continuous transformations of sets of ω-limit points,* Dop. AN URSR **11** (1965) 1407–1410 (Ukrainian with Russian and English summaries); MR **33** # 1847.

[7] *Behavior of mappings in a neighborhood of an attracting set,* Ukr. Matem. Zh. **18** (1966) 60–83 (Russian); Ref Zh M, 1966, 11 A 286.

[8] *On the set of convergence of one-dimensional iterations,* Dop. AN URSR 7 (1966) 866–870 (Ukrainian); Ref Zh M, 1966, 12 Б 78.

[9] *Continuous mappings on the set of limit points of an iterative sequence,* Ukr. Matem. Zh. **18** (1966) 127–129 (Russian); Ref Zh M, 1967, 2 Б 64.

[10] *Structure of an endomorphism on the ω-limit set,* Tezisy kratkikh nauchnykh soobshcheniĭ Mezhdunarodnogo kongressa matematikov, Moscow **6** (1966) 51 (Russian).

[11] *Partially ordered system of attracting sets,* Doklady AN SSSR **170** (1966) 1276–1278 (Russian); MR **35** # 311.

[12] *On a theorem of G. Birkhoff,* Dop. AN URSR, A **5** (1967) 429–432 (Ukrainian with Russian and English summaries); MR **35** # 3646.

[13] *Attracting sets containing no cycles,* Ukr. Matem. Zh. **20** (1968) 136–142 (Russian); MR **37** # 908.

153

Literature

SHCHERBAKOV, B. A.

[1] *The classification of Poisson stable motions and pseudorecurrent motions*, Doklady AN SSSR **146** (1962) 322–324 (Russian); MR **26** # 3247. English translation, Soviet Math. Doklady pp. 1320–1322.

[2] *On classes of Poisson stable motions. Pseudo-recurrent motions*, Izv. AN MSSR (Seriya estestv. i tekhn. nauk) **1** (1963) 58–72 (Russian with Moldavian summary); MR **35** # 998.

[3] *Decomposition of a set of Poisson stable motions into invariant sets*, Doklady AN SSSR **152** (1963) 71–74 (Russian); MR **27** # 3884.

[4] *Constituent classes of Poisson stable motions*, Sib. Matem. Zh. **5** (1964) 1397–1417 (Russian); MR **30** # 2201.

[5] *Recurrent functions and recurrent motions*, Izv. AN MSSR **7** (1965) 80–89 (Russian with Moldavian summary); MR **33** # 4405.

[6] *Recurrent solutions of differential equations*, Doklady AN SSSR **167** (1966) 1004–1007 (Russian); MR **34** # 4642. English translation, Soviet Math. Doklady **7** (1966) 534–538.

[7] *On a homeomorphic mapping of compact dynamical systems into the Bebutov dynamical system*, Izv. VUZov (Matematika) **12** (1967) 97–103 (Russian); MR **36** # 5926.

[8] *Recurrent solutions of differential equations and the general theory of dynamical systems*, Dif. Uravneniya **3** (1967) 1450–1460 (Russian); MR **36** # 6728.

*[9] *Poisson stable solutions of differential equations and topological dynamics*, Dif. Uravneniya **5** (1969) 2144–2155 (Russian); Ref Zh M, 1970, 4 Б358.

*[10] *Principle of coordinated restorability and Poisson stable solutions of differential equations*, Matem. Issled., RIO AN MSSR **2** (1970) 168–177 (Russian); MR **44** # 1881; Ref Zh M, 1971, 1 Б 380.

*[11] *The method of limiting transformations in the problem of the existence of Poisson stable solutions of differential equations*, Doklady Akad. Nauk SSSR **190** (1970) 796–799 (Russian); Ref Zh M, 1970, 6 Б 312.

*[12] *Topological dynamics and the Poisson stability of solutions of differential equations*, Kishenyov, Shtiintsa, 1972 (Russian); Ref Zh M, 1972, 6 Б 281 K.

SIBIRSKY, K. S.

[1] *Uniform approximation of points of dynamical limit sets and motions in them*, Doklady AN SSSR **146** (1962) 307–309 (Russian); MR **26** # 2698. English translation: Soviet Math. Doklady **3** (1963) 38–48.

[2] *Uniform approximation of points and properties of motions in dynamical limit sets*, Izv. AN MSSR (seriya estestv. i tekh. nauk) **1** (1963) 38–48 (Russian); Ref Zh M, 1964, 10 Б 172.

SIBIRSKY, K. S. and N. K. CHEBAN

*[1] *Classification of Poisson stable points in dispersive dynamical systems*, Doklady Akad. Nauk SSSR **209** (1973) 801–804 (Russian); English translation: Soviet Math. Doklady **14** (1973) 542–546; Ref Zh M, 1973, 8 Б 297.

154

SIBIRSKY, K. S. and I. A. CHIRKOVA
*[1] *Discontinuous dispersive dynamical systems*, Matem. Issled., RIO AN MSSR **4** (1971) 165–175 (Russian); Ref Zh M, 1972, 2 Б 312.
*[2] *On the theory of discontinuous dispersive dynamical systems*, Doklady Akad. Nauk SSSR **211** (1973) 1302–1305 (Russian); Ref Zh M, 1973, 12 Б 367.

SMALE, S.
[1] *Differentiable dynamical systems*, Bull. AMS, **73** (1967) 747–817; MR **37** # 3598.

STAKHI, A. M.
[1] *Classes of motions with periods in partially ordered dynamical systems*, Dif. Uravneniya, **1** (1965) 619–624 (Russian); Ref Zh M, 1965, 11 Б 221.
[2] *Topological properties of partially ordered dynamical systems*, Doklady AN SSSR, **172** (1967) 1032–1035 (Russian); MR **35** # 1000.
[3] *Invertibility of points and regions in partially ordered dynamical systems*, Dif. Uravneniya **4** (1968) 693–701 (Russian); Ref Zh M, 1968, 9 Б 218.
[4] *Minimal sets and Lyapunov stability in partially ordered dynamical systems*, Dif. Uravneniya **4** (1968) 1608–1615 (Russian); MR **38** # 5199.
[5] *Uniform approximation of sets and proximality in partially ordered dynamical systems*, Dif. Uravneniya **5** (1969) 1101–1106 (Russian); MR **39** # 7239.

STEPANOV, V. V.
[1] *Sur une extension du théorème ergodique*, Comp. Math. **3** (1936), 239–253.

SZEGŐ, G. P. and G. TRECCANI
*[1] *Semigruppi di transformationi multivoche*, Lecture Notes in Math., Springer-Verlag, Berlin-New York **101**, 1969; MR **40** # 6008.

URA, T.
*[1] *On the flow outside a closed invariant set; stability, relative stability and saddle sets*, Contributions to Diff. Equations **3** (1964) 249–294; MR **29** # 321.

URA, T. and I. KIMURA
*[1] *Stability in topological dynamics*, Proc. of the Japan Acad. **40** (1964) 703–706; MR **31** # 1669.

VINOGRAD, R. È.
[1] *On the limit behavior of an unbounded integral curve*, Uch. Zap. Moskovsk. Un-ta (Matematika) **155** (1952) 94–136 (Russian); MR **18** # 482.

VRKOČ, I.
[1] *The topological structure of the set of stable solutions of a differential system*, Chekhosl. Matem. Zh. **11** (1961) 262–305 (Russian summary); MR **23** # A 2593.
*[2] *Integral stability*, Chekhosl. Matem. Zh. **9** (1959) 71–129; MR **31** # 6031.

WILSON, F. W. and J. A. YORKE
*[1] *Lyapunov functions and isolating blocks*, J. of Diff. Equations **13** (1973) 106–123; Ref Zh M, 1973, 9 Б 327.

Literature

YOSHIZAWA, T.
  *[1] *Stability theory by Liapunov's second method,* Publ. of the Math. Soc. Japan, Tokyo, 1966; MR **34** # 7896; Ref Zh M, 1967, 9 Б 237 K.

ZHIDKOV, V. I.
  [1] *Some properties of discrete dynamical systems,* Uch. Zap. Moskovsk. Un-ta (Matematika) **163** (1952) 31–59 (Russian).

ZHIKOV, V. V.
  *[1] *On the problem of the existence of almost periodic solutions to differential and operator equations,* Sb. nauchn. trudov, Matematika, Vladimirsk. vech. politekhnich. in-ta (1969) 94–187 (Russian).

ZUBOV, V. I.
  [1] *Methods of Lyapunov and their application,* Leningrad, Izd-vo Leningradsk. Un-ta, 1957 (Russian); MR **19** # 275; Groningen, P. Noordhoff N.V., 1964; MR **31** # 3676.
  *[2] *Stability of motion (Methods of Lyapunov and their applications),* Vysshaya shkola, Moscow, 1973 (Russian); Ref Zh M, 1973, 272 K.

INDEX

Index

Index

M

Index